I0393796

NATIONAL SPACE WEATHER PROGRAM

Implementation Plan
2nd Edition

Prepared by the Committee for Space Weather

for the

National Space Weather Program Council

Office of the Federal Coordinator for Meteorology

FCM-P31-2000
Washington, DC
July 2000

FOREWORD

We are pleased to present this Second Edition of the *National Space Weather Program Implementation Plan*. We published the program's *Strategic Plan* in 1995 and the first *Implementation Plan* in 1997. In the intervening period, we have made tremendous progress toward our goals but much work remains to be accomplished to achieve our ultimate goal of providing the space weather observations, forecasts, and warnings needed by our Nation. Over the course of the last two years, and in conjunction with the Space Weather Architecture Study conducted by the National Security Space Architect, the interagency Committee for Space Weather prepared this Plan. It provides a wealth of information on the progress made to date as well as refining guidance for the direction of future activity. We look forward to even greater success in the next three years.

The National Space Weather Program Council members whose signatures appear below are committed to this program and will work toward its implementation through agency programs. This updated Implementation Plan sets the direction for the research, operations, education, training, and program management to achieve the goals of the National Space Weather Program.

NATIONAL SPACE WEATHER PROGRAM COUNCIL

Mr. Samuel P. Williamson, Chairman
Federal Coordinator

Dr. David L. Evans
Department of Commerce

Colonel Michael A. Neyland, USAF
Department of Defense

Dr. Margaret S. Leinen
National Science Foundation

Dr. Edward J. Weiler
National Aeronautics and Space
Administration

Mr. Robert E. Waldron
Department of Energy

Mr. David Whatley
Department of Transportation

Mr. James F. Devine
Department of the Interior

TABLE OF CONTENTS

LIST OF FIGURES

LIST OF TABLES

Table   **Page**

EXECUTIVE SUMMARY

In the six years since the inception of the National Space Weather Program (NSWP), space weather has virtually become a household word. Space weather refers to conditions on the Sun and in the solar wind, magnetosphere, ionosphere, and thermosphere that can influence the performance and reliability of space-borne and ground-based technological systems and can endanger human life or health. The growing awareness within the general public is largely the result of a conscious effort by all space weather stakeholders to educate the nation on the effects of solar storms and their potential impacts on the modern technology so important to daily life. In addition to increased awareness, solid advances have been made in our knowledge of the space weather system and in our ability to forecast potentially disruptive space weather events.

The NSWP *Strategic Plan*, released in 1995, put forth a strategy for achieving space weather goals. The program elements outlined in this strategy are shown in the figure below. The *Strategic Plan* was followed in 1997 by the NSWP *Implementation Plan*, which identified specific objectives and recommended activities necessary for improving space weather predictive capabilities. In the last six years, significant progress has been made in all programmatic areas. The NSWP's success is a result of the concerted efforts by the government agencies actively involved in space weather, as well as by all

National Space Weather Program Elements

Figure ES-1. National Space Weather Program Elements

stakeholders in government, academia, and industry. The NSWP has provided a context for joint programs, improved communication, and effective sharing of information and results.

The National Oceanic and Atmospheric Administration's Space Environment Center and the U.S. Air Force's 55th Space Weather Squadron constitute the operational arms of the nation's space weather support capabilities. By making use of space-based and ground-based sensors and state-of-the-art computer models, the operations centers provide support for civilian and military customers and systems. Despite the progress made in the last six years, current capabilities still fall short of requirements for warning, nowcasting, forecasting, and post-analysis, although post-analysis capabilities are the most robust. In many areas, significant shortfalls remain and much work remains to be done.

NSWP agencies partnered to create a convenient set of metrics to better quantify the progress being made. Because of the many scientific and technical disciplines involved in the NSWP, developing such metrics is a difficult task. For this reason, a panel of scientists convened to formulate metrics in the three space weather domains. The domains are the ionosphere/thermosphere, the magnetosphere, and solar/solar wind. These research metrics are designed to assess the fundamental understanding of space weather processes. They will be adopted by the space science community and tracked over the next decade as an indicator of scientific progress as well as to identify the most serious gaps in capability.

The National Science Foundation (NSF), National Aeronautics and Space Administration (NASA), National Oceanic and Atmospheric Administration (NOAA), and the Department of Defense are supporting aggressive research programs striving to achieve space weather goals. In addition to research efforts aimed at specific goals, many areas of research in space and plasma physics indirectly support program objectives by advancing knowledge in fundamental scientific areas. However, participants recognized early in the program that rapid progress could be made only by implementing a more targeted research program. Toward this end, NSF, with contributions from the Air Force Office of Scientific Research and the Office of Naval Research, began funding competitively-selected research proposals in key areas. They held these competitions in 1996, 1997, 1999, and 2000. The program announcements included a description of the areas of scientific emphasis to fill gaps in our existing knowledge and predictive capabilities. Panels of experts selected proposals based on the merit of the research and their potential to contribute to space weather goals. Results from grants awarded under these programs have included many innovative approaches to space weather model development and predictive capabilities, as well as more basic research aimed at improving our understanding of space weather phenomena. Overall, more than 70 targeted space weather proposals have been funded.

In recent years, models supporting space weather research have improved tremendously. Model development has emphasized physics-based approaches that are more likely to lead to improved predictive capabilities. Adaptive grids and data assimilation techniques have been incorporated into some of the larger and more complex magnetohydrodynamic models describing solar processes and solar wind-magnetosphere interactions. Emphasis has also been placed on combining models to create a more seamless description of the entire space weather system from the Sun to Earth's atmosphere.

The use of research observations to support space weather priorities has also blossomed during the last six years. Many ongoing and planned space-based missions have been developed with space weather requirements in mind, making it easier to adapt the data flow for operational purposes. The explosion of the World Wide Web has also facilitated the faster and broader sharing of space- and ground-based measurements for space weather purposes. Enabled by the NSWP, the excellent coordination between the producers of these research observations and the operational

One interagency success story…

The success of solar storm readiness is dependent on collaboration across the members of the National Space Weather Program. This was evident during the major geomagnetic storm that occurred from May 4-8, 1998. During the previous week forecasters and scientists mobilized into a high state of readiness as active regions on the sun produced a series of intense solar flares. NOAA forecasters monitoring the activity received numerous calls from the SOHO science team confirming that several large coronal mass ejections directed towards Earth had also occurred. NOAA forecasters used this information to estimate when they should expect to see the disturbances pass ACE. The storms passed ACE as projected and forecasters were able to alert customers with approximately 40 minutes of lead time. In response to the forecast and their own data, electric utilities cut power import from Canada by half and increased safety margins on other parts of the grid.

The storm was estimated to be about one quarter of the magnitude of the March 13, 1989 storm, but was still serious enough to cause great concern and stress on power and other systems affected by geomagnetic activity. The cooperation between NOAA, NASA, USAF, NSF and other NSWP partners was critical to preventing major system damage and failures. If one link in the NSWP chain had broken, the outcome may have been vastly different.

community has produced new opportunities for collaboration. Of particular note is NASA's Living with a Star initiative which will provide observational capabilities, theory, and modeling to enhance knowledge of the space weather system.

Working together, the stakeholding agencies of the federal government have developed new timelines for both operational and research-level models and sensors. These timelines include critical milestones in model development and sensor deployment to guarantee improved space weather forecasting capabilities. Near-term emphasis in NSWP research includes the study of the origins of coronal mass ejections, the triggering of magnetospheric substorms, and the evolution of ionospheric irregularities. For space weather modeling, validation and testing of existing models along with the application of data assimilation techniques are high priorities. The emphasis for observing is to maintain the existing suite of ground- and space-based observatories, take advantage of real-time data when possible, and plan for future missions and facilities, such as NSF's planned Relocatable Atmospheric Observatory, to address gaps in knowledge and observational coverage.

The transfer of research knowledge into operations, commonly known as technology transition, is an area that has long impeded progress in applying scientific research results to operational needs within the space weather forecasting communities. As a result, NSWP agencies initiated a multi-faceted approach to address this issue. Both NOAA and the DOD implemented plans to develop Rapid Prototyping Centers (RPCs) whose function is to adapt space weather models for use in the operational centers. To aid researchers in the development of models, the DOD, NASA, NSF, and NOAA have worked together to create the Community Coordinated Modeling Center (CCMC). A parallel-processing supercomputer at Air Force Weather Agency in Omaha, Nebraska, is linked to front-end workstations at Goddard Space Flight Center, Maryland. The facility provides researchers with a means to exercise models that may eventually be used at the operational centers. The CCMC is intended to provide the critical link between the scientific research community and the Rapid Prototyping Centers as well as establishing data management standards and procedures early in development (see Figure ES-2).

Educational activities play an important role in space weather by elevating awareness of potential impacts, both among space weather customers and the general public. NSWP participants have used both formal and informal venues for educational and outreach efforts in space weather. Educational materials on space weather are being made web-accessible to schools at the K-12 level, while NOAA is producing a curriculum guide on "Solar Physics and Terrestrial Effects" for teachers of grades 7 through 12. Awareness in the general public has been enhanced through the proliferation of space weather web sites and the extensive media coverage in the past several years. Space scientists have made a concerted effort to publicize space weather events and their effects on technical systems. They have participated in the design and development of web sites through which the development of events can be observed in real time. The net result of these efforts has been an unprecedented level of attention to space weather by the general public. Specific

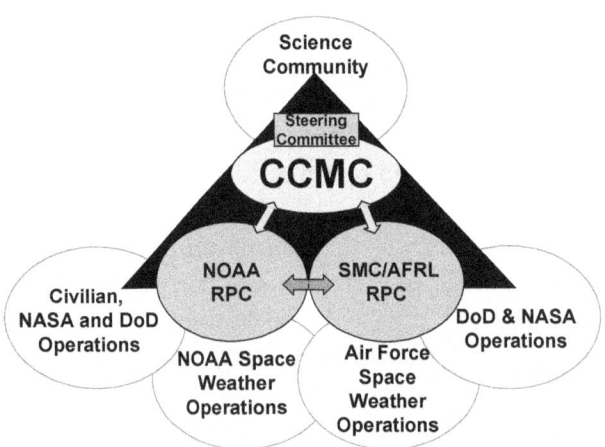

Figure ES-2. Linking the Science Community with Operational Centers

 examples include a television program on space weather on the Discovery Channel™, and an IMAX™ film on solar maximum. Press coverage, inspired by carefully planned press releases, has increased tremendously as indicated by the number of articles appearing in newspapers and magazines.

Equally important for effective progress in space weather goals is to establish and strengthen relationships with space weather customers. Interaction between space weather customers and operational forecasters had previously taken place at the Space Environment Center in Boulder during the Space Weather User Conferences held every three years. In 1998, meeting attendees favored meeting annually, at least during the solar maximum period, and also endorsed merging the meeting with the Space Weather Research to Operations Workshops previously held in January in Boulder. The two meetings were held consecutively for the first time in April 1999 in an event referred to as Space Weather Week. This successful format for encouraging feedback between the scientific community and space weather customers will continue on an annual basis for the next several years.

In addition to the Users Conference, program participants conducted two other more specialized workshops. The first was a workshop on Geomagnetically Induced Currents held at the Electric Power Research Institute (EPRI) headquarters in Washington, DC, in October 1996. The second was a workshop on Space Weather Effects on Navigation and Communication Systems held at COMSAT headquarters in Bethesda, MD, in 1997.

Customer education will continue to have a high priority in the NSWP and is expected to help in collecting and analyzing space weather impacts on operations. It is also expected to improve the identification of customer requirements.

SOHO recovered…

On June 25, 1998 NASA radio contact with SOHO was suddenly lost, disrupting a tremendously successful science mission and terminating a valuable source of data on Coronal Mass Ejections for NOAA's Space Environment Center. The effort to find SOHO and recover the spacecraft brought experts from various NSWP agencies together in a common mission. With the encouragement of a NOAA/SEC scientist, staff at NSF's National Astronomy and Ionospheric Center used the Arecibo radio telescope to transmit a signal toward SOHO's last known location on July 23, 1998. Using radar techniques NASA's DSN groundstation at Goldstone received the signal containing SOHO's echo and tracked the spacecraft for more than an hour. The radar data indicated that SOHO was still in its proper orbit, but its solar panels were pointing away from the sun and it had begun slowly turning.

Thanks to the team effort of National Space Weather Program partners and their colleagues SOHO was gradually brought back into service. SOHO first responded to radio transmissions on August 3, and telemetry from SOHO was received August 8, telling controllers the condition of the spacecraft and its instruments. The spacecraft's frozen hydrazine fuel was gradually thawed, and on September 16, SOHO's thrusters were fired to stop its spin and to place it in the correct orientation towards the Sun. Over the next few weeks SOHO's instruments were gradually turned on and tested culminating in mid-October with the release of the first images.

Identifying the requirements of the satellite industry has been difficult due to the highly competitive nature of the industry and the complexities associated with insurance coverage and the legal aspects of satellite communications. To initiate discussions with satellite industry representatives, NSF made an award for the purpose of interviewing top aerospace industry executives. This study underscored the difficulties in addressing commercial space weather needs. The DOD has also initiated efforts with each of its services to improve and expand the documentation of space weather impacts on military operations.

Program management for the NSWP continues to be administered through the Office of the Federal Coordinator for Meteorology (OFCM). The National Space Weather Program Council provides top-level direction, while day to day oversight of the program is entrusted to the Committee for Space Weather aligned under the Council. Parallel efforts in management and coordination of space weather activities have been launched by the DOD through the office of the National Security Space Architect and by NASA through its Living with a Star initiative. The National Security Space Architect has concluded a detailed space weather architecture study and developed a transition plan to meet the nation's space weather needs through the year 2025. Within the framework of the NSWP, the architecture and transition plan provide detailed guidance for DOD agencies to use in planning, funding, developing, acquiring, and operating its portion of the overall architecture. Because of the interagency nature of the NSWP and the nation's space weather operations, many of the NSWP agencies participated in the project and the results have been incorporated in NSWP planning. Similarly, mission planning for Living with a Star is proceeding with close coordination between NASA and other NSWP agencies.

International efforts in space weather have also been greatly expanded as a result of the increasing U. S. activities. The importance of international participation has been continually stressed within the NSWP because it provides a means to blend the unique resources of other nations into the space weather planning process. The effects of space weather are global; experimental campaigns in support of research, as well as environmental monitoring, include the assets that foreign scientists and observing capabilities can contribute. Among the other nations with dedicated space weather programs are Sweden, Australia, Japan, Canada, France, Taiwan, and China. The past few years have seen a tremendous improvement in international coordinating activities, particularly through the Scientific Committee for Solar Terrestrial Research. Among its accomplishments, the S-RAMP Steering Committee under SCOSTEP identified a period of high solar activity as a "Special Study Interval" for space weather, and organized a dedicated campaign in September of 1999.

In conclusion, the past six years have seen significant progress made in all programmatic areas demanded by space weather goals. The success of these endeavors is directly attributed to the excellent cooperation among government agencies and space weather stakeholders. With this spirit of cooperation, optimism for future success remains high.

CHAPTER 1

INTRODUCTION

As we begin the new millennium, our civilization's reliance on technology affected in some way by space weather continues to grow at a rapid pace. Space weather refers to conditions on the Sun and in the solar wind, magnetosphere, ionosphere, and thermosphere that can influence the performance and reliability of space-borne and ground-based technological systems and can endanger human life or health. The National Space Weather Program (NSWP) emerged in 1994 from the efforts of several U.S. government agencies to prepare us to deal with the vulnerabilities of our technology. Through the Office of the Federal Coordinator for Meteorological Services and Supporting Research (OFCM), these agencies documented the goals of that program in the *National Space Weather Program Strategic Plan* (FCM-P30-1995, Office of the Federal Coordinator for Meteorological Services and Supporting Research, Silver Spring, MD, 1995). The interagency focus and cooperation continued in 1997 with the development of the *National Space Weather Program Implementation Plan* (FCM-P31-1997, Office of the Federal Coordinator for Meteorological Services and Supporting Research, Silver Spring, MD, 1997) which provided more specific direction to the federal government's space weather efforts.

This *Implementation Plan* is a living document and this is its first revision. It has been developed concurrently with the National Security Space Architect's Space Weather Architecture and describes the linkage to and incorporation of that architecture into the National Space Weather Program. In this revision, we build on the previous plan and report on the significant accomplishments in research, operations, technology transition, education and outreach. We have updated the program's timelines and offer specific recommendations to carry us forward. This is the culmination of months of multi-agency coordination and cooperation and represents a dedicated effort by Federal agencies to improve capabilities in an area with critical societal impacts.

1.1 History of the Program

In 1993, members of the space science community visited the National Science Foundation (NSF) and raised the issue of improving the Nation's ability to specify and forecast space weather. In response, NSF organized a meeting at which representatives from government, industry, and academia met and discussed the current status of space weather research and operational systems. These discussions highlighted deficits in current capabilities and suggested that much could be gained by better coordination of efforts across Federal agencies. Other issues presented at the meeting included the limitations imposed by budget constraints and the reluctance of industry to reveal problems with current systems. It became clear that an overarching program to coordinate space weather activities would help to more effectively apply limited resources, and that such a program should be overseen by an organization composed of only government agencies so that industrial participants would be more willing to identify problems with existing systems.

Because the Office of the Federal Coordinator for Meteorological Services and Supporting Research, more briefly known as the Office of the Federal Coordinator for Meteorology (OFCM), orchestrates multi-agency coordination, it became the focal point for developing the space weather program. Furthermore, the Committee for Space Environment Forecasting (CSEF) already existed within the OFCM structure. Planning for a space weather program represented a logical extension of the responsibilities of that committee. CSEF appointed the Working Group for the National Space Weather Program (WG/NSWP) and charged it with developing a strategic plan. Work began in the summer of 1994 involving representatives from NSF, Department of Defense (DOD), National Oceanic and Atmospheric Administration (NOAA), National Aeronautics and Space Administration (NASA), Department of Energy (DOE), and Department of the Interior (DOI). In parallel with these efforts, the Working Group recommended that a Program Council be created to provide high-level, multi-agency oversight for the emerging program. The Working Group developed a draft charter for the Program Council and on August 4, 1995, the Program Council met for the first time. They adopted the charter and approved the Strategic Plan. The Program Council then directed the Working Group to develop an implementation plan for the NSWP. This process began with the identification of customer requirements for space weather services. Once these requirements were specified, representatives from the research community met to develop a road map laying out research, modeling, and observational requirements that would lead to achievement of the goals. The Program Council approved the initial Implementation Plan in January 1997. The present document is the first update of that initial plan.

1.2 Scope of the Program

The NSWP encompasses all activities necessary for the timely specification and forecast of natural conditions in the space environment that may have an impact on technical systems and human life or health. The domains of primary interest to the program include the Sun and solar wind, the magnetosphere, the ionosphere, and the thermosphere. Because of the vastness and complexity of the region of interest, all traditional areas of space sciences can contribute to achieving the program goals.

Space weather begins at the Sun's surface, the source of radiative and particle energy impacting Earth. Solar activity changes the radiative and particle output of the Sun, producing corresponding changes in the near-Earth space environment, as well as at Earth's surface. The most dramatic events on the Sun, insofar as space weather effects are concerned, are solar flares and coronal mass ejections. Although longer term variations in solar emissions do not produce dramatic space weather effects, they are important in helping us understand the underlying processes behind the short-term variations.

Changes in the radiative output from the Sun directly affect the state of the upper atmosphere and ionosphere through the excitation and ionization of atoms and molecules. Particle emissions from the Sun include both the energetic particles and the low-energy plasma that constitute the solar wind. Both particles and electromagnetic fields evolve as they flow outward from the Sun, especially as they create or interact with interplanetary shocks.

The solar wind moves outward from the Sun and impinges on Earth. The plasma and magnetic field of the solar wind interact with Earth's atmosphere and geomagnetic field, creating a tear-drop-shaped region called the magnetosphere. The surface of this region is referred to as the magnetopause. The magnetopause is usually found near 10 Earth radii (R_E) in the sunward direction, although this distance is highly variable (roughly between 5 and 15 R_E) in response to solar wind dynamic pressure. In the antisunward direction, the magnetopause extends to distances beyond the orbit of the moon. The magnetopause represents a barrier that prevents all but a fraction of the energy carried by the solar wind from entering the magnetosphere. Under normal conditions, the energy that does penetrate the magnetopause is stored in the form of the particles and fields of the magnetosphere, but under some conditions it is impulsively released into Earth's atmosphere. This impulsive release of energy is referred to as a magnetospheric substorm. It is characterized by the appearance of bright, dynamic aurora and the development of intense ionospheric currents. During a substorm the magnetic field in the magnetosphere suddenly assumes a new configuration; after the substorm there is a recovery period that takes many hours.

Substorms are a relatively short-lived magnetospheric response to solar wind stimulus. Geomagnetic storms are a sustained, long-lived (days to weeks) response to a prolonged period of solar wind flow characterized by a strong southward interplanetary magnetic field. Geomagnetic storms lead to a substantial energization of the ring current, a belt of quasi-trapped electrons, protons, and heavier ions, as well as significant geomagnetic fluctuations at low geographic latitudes. Magnetospheric particles precipitate into the polar caps, heating the neutral atmosphere (thermosphere and mesosphere) and launching ionospheric disturbances. Substorms may also occur during the course of geomagnetic storms. Once the solar wind returns to its undisturbed state, the magnetosphere and ionosphere require hours to days to recover.

Because Earth's magnetic field permeates the magnetosphere, most magnetospheric processes are manifested in some way by changes in the properties of the ionosphere and thermosphere. Magnetospheric processes produce electrical currents, auroral emissions, frictional heating, ionization, and scintillation. All of these phenomena are elements of near-Earth space weather. The near-Earth space environment is also influenced by processes originating at lower altitudes, such as gravity waves, and direct energy deposition from solar radiation and cosmic rays. Space weather effects also include the electrical currents induced within Earth's surface as a result of changes in ionospheric currents.

This brief description of the space weather system demonstrates the vastness of the region of interest to the NSWP and the complexity of the physical processes that must be understood. Adding to this complexity is the high degree of coupling between the various regions. The program will emphasize the importance of dealing with the space environment as a seamless system in which processes occurring in one location cannot be understood without adequate knowledge of the way the entire system is linked.

The NSWP is primarily concerned with naturally occurring phenomena in the space environment but addresses Department of Defense concerns with man-made space environmental effects. Although the program does not specifically address the possible impact of orbital debris on satellite systems, it will contribute to the accurate tracking of objects in space by improving the specification and prediction of variations in atmospheric density, which affect the drag on orbiting objects.

Similarly, the NSWP does not deal directly with the engineering aspects that enter into the design and development of technical systems. Here again, the program can be of benefit to the community by providing detailed information about the space environment so that engineers can better design these systems. Often accurate specification of the range in environmental parameters to which a piece of equipment will be subjected can result in significant cost savings.

The goal of the NSWP is to provide products to a community of customers that is continually changing. Each of these customers may have different requirements, making it a formidable task to provide customized products. The routine production of information tailored to meet specific customer requirements is not within the scope of the NSWP. The specification and forecast information provided by the forecast centers will be sufficient to allow such tailored products to be developed either by the customers themselves or by others offering to provide these services. In particular, the need for these services provides opportunities for small businesses or other profit-making enterprises. In the case of Department of Defense (DOD) customers, DOD takes responsibility for its own tailored products. Agency representatives and customers involved in the NSWP will routinely evaluate the status of NSWP products and agree upon the level of information falling within the scope of the program.

1.3 Relevance to the Nation

Space weather is working its way into the national consciousness as we see an increasing number of problems with parts of our technological infrastructure, such as satellite disruptions and failures as well as electric power brownouts and blackouts. As our society grows more dependent on advanced technology systems, we become increasingly more vulnerable to malfunctions in those systems.

For example, electrical power networks connecting widely separated geographic areas have increased the probability of power grids absorbing damaging electric currents induced by geomagnetic storms. The miniaturization of electronic components and reduced radiation hardening on satellites makes them potentially more susceptible to damage by high-energy particles. Similarly, aircraft designed to fly at 60,000 feet (18.3 kilometers) have increased human risk to radiation exposure during severe space weather. Figure 1-1 lists sample significant space weather events and their impacts over the past several years.

System and human vulnerabilities to space weather effects include the following:

Engineering Aspects. Engineers use space environment information to specify the extent and types of protective measures that are to be designed into a system and to develop operating plans that minimize space weather effects. However, engineering solutions to some problems may be very costly or impossible to implement. After the fact, engineers use space environment information to determine the source of failures and develop corrective actions. Significant economic and societal benefits can be realized if designers of emerging technology can (1) anticipate the properties of the space environment to which the hardware will be subjected, (2) depend on accurate and timely predictions of space weather, and (3) take advantage of post-event analysis to determine the source of system anomalies and failures and to build a database for future planning.

Satellite Systems. Space weather affects satellite missions in a variety of ways, depending on the orbit and satellite function. Our society depends on satellites for weather information, commercial television, communications, navigation, exploration, search and rescue, research, and national defense. The impact of satellite system failures is more far-reaching than ever before, and the trend will almost certainly continue at an increasing rate.

Energetic particles that originate from the Sun, from interplanetary space, and from Earth's magnetosphere continually impact the surfaces of spacecraft. Highly energetic particles penetrate electronic components, causing changes in electronic signals that can result in spurious commands within the spacecraft or erroneous data from an instrument. These spurious commands have caused major satellite system failures that might have been avoided if ground controllers had had advance notice of impending particle hazards. Less energetic particles contribute to a variety of spacecraft surface charging problems, especially during periods of high geomagnetic activity. In addition, energetic electrons responsible for deep dielectric charging can degrade the useful lifetime of internal components. Overall radiation dose can ultimately determine satellite lifetime.

Highly variable solar ultraviolet radiation continuously modifies terrestrial atmospheric density and temperature, affecting spacecraft orbits and lifetimes. Major geomagnetic storms result in heating and expansion of the atmosphere, causing significant perturbations in low-altitude satellite trajectories. At times, these effects may be severe enough to cause premature re-entry of orbiting objects, such as Skylab in 1979. It is important that satellite controllers be warned of these changes and that accurate models be in place to realistically predict the resulting atmospheric effects. The Space Shuttle is also vulnerable to changes in atmospheric drag; re-entry calculations for the orbiter are highly sensitive to atmospheric density, and errors can threaten the safety of the vehicle and its crew.

Power Systems. Modern power grids are extremely complex and widespread and potential changes in the industry will increase the interconnection of regional grids. The long power lines that traverse the Nation are susceptible to electric currents induced by the dramatic changes in high-altitude ionospheric currents that occur during geomagnetic storms. "Surges" in power lines from induced currents can cause massive network failures and permanent damage to multimillion-dollar equipment in power generation plants. Considering the significant national dependence on reliable electrical power, the resulting social chaos, economic impact, and threat to safety during widespread power outages are far more serious than the simple cost of repairing the systems.

The electric power distribution system has developed an increased susceptibility to the phenomenon of geomagnetically induced currents because of widespread grid interconnections, complex electronic controls and technologies, and large inter-area power transfers. The phenomenon occurs globally and simultaneously, and industry operations allow for little redundancy or operating margin to absorb the effects. Mitigation of such effects is fairly straightforward provided advance notice is given of an impending storm; specific strategies currently exist within the power industry. Advanced warnings of storms are needed, but of equal economic importance to industry is that the forecasts be reliable. False alarms are counterproductive and must be minimized.

Navigation Systems. The accuracy of maritime navigation systems using very low frequency signals, such as Long-Range Navigation (LORAN), depends on knowing accurately the altitude of the bottom of the ionosphere. Rapid vertical changes in this boundary during solar flares and geomagnetic storms can introduce errors of several kilometers in location determinations.

The Global Positioning System (GPS) operates by transmitting radio waves from satellites to receivers on the ground, aircraft, or other satellites. These radio signals are used to calculate location very accurately. However, significant errors in positioning can result when the signals are refracted and slowed by ionospheric conditions significantly different from normal. In addition, receivers can experience loss of GPS signal lock when the signal traverses an ionospheric disturbance (scintillation). Future high-resolution applications of GPS technology will require better space weather support to compensate for these induced errors. Accurate specification and prediction of the properties of the ionosphere will aid in the design and operation of emerging systems.

March 24, 1940. A "great" geomagnetic storm rendered inoperative 80% of all long-distance telephone connections out of Minneapolis, Minnesota. Electric service was temporarily disrupted in portions of New England, New York, Pennsylvania, Minnesota, Quebec, and Ontario.

February 9-10, 1958. A geomagnetic storm caused severe interruptions on Western Union's North Atlantic telegraph cables and made voice communications very difficult on the Bell System transatlantic cable from Newfoundland to Scotland. Toronto, Canada, experienced a temporary blackout.

August 4, 1972. A severe geomagnetic storm caused a 30-minute shutdown of the Bell System coaxial cable link between Plano, Illinois, and Cascade, Iowa. A power transformer failed at the British Columbia Hydro and Power Authority.

November 26, 1982. The Geostationary Operational Environmental Satellite (GOES) 4 visible and infrared spin-scan radiometer, which maps cloud cover, failed 45 minutes after the arrival of high-energy protons from a major solar flare. The untimely failure occurred as a series of intense storms hit the California coast.

March 13-14, 1989. A severe geomagnetic storm caused a system-wide power failure in Quebec, Canada, resulting in the loss of over 20,000 megawatts. The blackout cut electric power to several million people. Time from onset of problems to system collapse was about 90 seconds. High frequency (HF) radio frequencies were virtually unusable worldwide, while very high frequency (VHF) transmissions traveled unusually long distances and created interference problems. A Japanese communications satellite lost half of its dual-redundant command circuitry. A National Aeronautics and Space Administration (NASA) satellite dropped 3 miles (4.8 kilometers) in its orbit due to the increase in atmospheric drag.

January - March, 1991. Coalition military forces experienced occasional high frequency radio communications interruptions due to scintillation.

April 29, 1991. A transformer at the Maine Yankee Nuclear Plant catastrophically failed within a few hours of a severe geomagnetic storm onset.

January 20-21, 1994. Two Canadian communications satellites failed, interrupting telephone, television, and radio service for several hours. The failures occurred after an extended period of high electron levels in the satellite environment.

May 1998. Solar activity may have been the cause of a significant disruption of pager service across the United States when the Galaxy 4 satellite experienced problems.

Figure 1-1. Impacts of Significant Space Weather Events

Communications. Radio communications over a broad range of frequencies are affected by space weather. High Frequency (HF) radio wave communication is more routinely affected because this frequency depends on reflection from the ionosphere to carry signals great distances. Ionospheric irregularities contribute to signal fading; highly disturbed conditions, usually near the aurora and across the polar cap, can absorb the signal completely and make HF radio propagation impossible. Accurate forecasts of these effects can give operators more time to find an alternative means of communication. Telecommunication companies increasingly depend on higher frequency radio waves, such as ultrahigh frequency (UHF), which penetrate the ionosphere and are relayed via satellite to other locations. Signal properties can be changed by ionospheric conditions so that they can no longer be accurately received at Earth's surface. This may cause degradation of signals, but more important, can prohibit critical communications, such as those used in search and rescue efforts and military operations.

Manned Space Flight. Besides being a threat to satellite systems, energetic particles present a hazard to astronauts on space missions. On Earth we are protected from these particles by geomagnetic shielding and the atmosphere. The geomagnetic field shields Earth's atmosphere from all particles of millions of electron-Volt (MeV) energy except in the polar regions. The atmosphere absorbs all but the most energetic cosmic ray particles. During space missions, astronauts performing extra-vehicular activities are relatively unprotected at high latitudes in the spacecraft orbit. This could be particularly problematic during the ongoing construction of the International Space Station during solar maximum. The fluxes of energetic particles can increase hundreds of times following an intense solar flare or to dangerous levels during a large geomagnetic storm. Timely warnings are essential to give astronauts sufficient time to return to their spacecraft prior to the arrival of such energetic particles. High altitude aircraft crews and passengers on polar routes, e.g., on supersonic transports (SSTs) or U-2s, are also susceptible to radiation hazards during similar events.

1.4 Summary of the Strategic Plan

Recognizing the need for a more coordinated effort to improve present capabilities in specifying and forecasting conditions in the space environment, Federal agencies representing the research, operations, and user communities initiated the NSWP, as outlined in *The National Space Weather Program Strategic Plan.* The overarching goal of the program is to achieve an active, synergistic, interagency system to provide timely, accurate, and reliable space weather warnings, observations, specifications, and forecasts within the next 10 years. By building on existing capabilities and establishing an aggressive, coordinated process to set national priorities, focus agency efforts, and leverage resources, the NSWP provides the path to attain this goal. The activities that the NSWP will conduct are listed in Figure 1-2 and the specific goals of the program are enumerated in Figure 1-3.

National Space Weather Program Activities

- Assess and document the impacts of space weather
- Identify customer needs
- Set priorities
- Determine agency roles
- Coordinate interagency efforts and resources
- Ensure exchange of information and plans
- Encourage and focus research
- Facilitate transition of research results into operations
- Foster education of customers and the public

Figure 1-2. National Space Weather Program Activities

The key elements of the NSWP are described as follows:

Forecast and Specification Services. The predominant driver of the program is the value of space weather forecasting services to the Nation. The accuracy, reliability, and timeliness of space weather specification and forecasting must become comparable to that of conventional weather forecasting. Early warning capabilities of impending dangerous conditions must become equally reliable to be valuable for mitigation purposes. The strengthening of services includes modernization of facilities; implementation of new models and other analysis and forecast techniques; improved education and training; improved production, design, and dissemination of forecast products; and improved communication with the users of the services. Proposed operational models, instrumentation, and techniques are evaluated according to their potential to improve forecasting services.

Research. This includes ongoing, intensive efforts to understand the fundamental physical processes that affect the state of the Sun, solar wind, magnetosphere, ionosphere, and atmosphere, with a focus on resolving research problems that impede improvements in forecasting capability. Radiative, dynamical, electrical, and chemical coupling between different regions have been and will continue to be studied using data from existing ground- and space-based instrumentation. Theoretical investigations in these areas will help to define the needed observations and will aid the development of operational models.

Observations. The program builds on existing observational capabilities and determines the value of current data and new data needs. Observations in support of research and forecasting are growing as critical parameters for forecasting are identified, measurement techniques are defined, and new space- and ground-based platforms are developed. The initial focus is on better coverage of data-void or data-sparse regions, and on the

---

The National Space Weather Program Goals

To advance
- observing capabilities
- fundamental understanding of processes
- numerical modeling
- data processing and analysis
- transition of research into operational techniques and algorithms
- forecasting accuracy and reliability
- space weather products and services
- education on space weather

To prevent or mitigate
- under- or over-design of technical systems
- regional blackouts of power utilities
- early demise of multi-million dollar satellites
- disruption of communications by satellite, HF, and VHF radio
- disruption of long-line communications
- errors in navigation systems
- excessive radiation doses, dangerous to human health

Figure 1-3. National Space Weather Program Goals

deployment of systems that provide data with appropriate accuracy, resolution, and timeliness. Because instrumentation development is an evolutionary process, the program emphasizes rapid exploitation of observational capabilities, bridging gaps between research observing systems and subsequent operational observing systems, and efficient communication between data analysts, researchers, and instrument designers.

Modeling. These efforts for specifying and predicting the space environment have been under way for several years and some operational benefits have been realized. The program continues to coordinate modeling and integration activities to ensure the consistency and optimal performance of the models. A primary goal is to develop physics-based specification and forecast models covering the forecast period out to 72 hours for solar events and 48 hours for near-Earth space weather phenomena. These models are evaluated in close collaboration with research and observation efforts and with regard to user requirements. Gaps and deficiencies in these models are identified and used to set requirements for future models.

Education. The education activities supported by the program enhance public awareness of space weather and its impacts; help ensure a sufficient supply of educated scientists and engineers to maintain expertise in all space-weather-related fields; and improve training of forecasters, observers, and system operators. An educated public and commercial sector are better able to utilize space environment forecasting services; student research will supply fresh ideas to explore; and knowledgeable government officials and the media will help realize the socioeconomic benefits.

Technology Transition and Integration. Although significant strides have been made through the Community Coordinated Modeling Center and the Rapid Prototyping Center, these processes must be continually improved to facilitate the transfer of tools, techniques, and knowledge from the research or commercial communities to the operational forecasting activities. This effort, often a bottleneck, is critical to the success of the program. Innovative means must be explored to nurture a dynamic process for technology exploitation and transition to improve forecasting capability, utilize all relevant research, and rapidly realize benefits.

1.5 The Implementation Plan

The preceding paragraphs describe space weather's range of impacts on our society and the context of the National Space Weather Program's overarching objectives. The remainder of this document reviews significant progress made since the original *Implementation Plan* was published in 1997 and reiterates and refines the means by which the various Federal agencies will achieve the program's objectives. The individual agencies are responsible for appropriate related actions under the Government Performance Results Act. The Act will not be specifically addressed by this document. The next chapter describes current capabilities and updates the Program's goals and strategy.

CHAPTER 2

CAPABILITIES, GOALS, AND STRATEGY

2.1 Background

Space weather services are currently provided by the National Oceanic and Atmospheric Administration (NOAA) Space Environment Center (SEC) and the United States Air Force (USAF) 55th Space Weather Squadron (55 SWXS), both in Colorado. The former serves civilian customers, while the latter serves the needs of the Department of Defense (DOD). At these centers, forecasting space weather is approached in much the same way as forecasting tropospheric weather. Data are collected, checked for quality, analyzed, fed into models, and displayed graphically. Forecasters mentally integrate numerical products, images, and other analyses, using experience, physical understanding, and empirical tools. A "most likely" scenario emerges of what the forecaster believes the state of the environment will evolve to from its present state. If a significant event (severe weather) is forecast, it must be assigned onset time, intensity, duration, and, if possible, how it will affect specific regions of the environment.

In assessing present capabilities we distinguish between four different types of space weather products--warnings, nowcasts, forecasts, and post-analysis--as follows:

A *warning* is given for an event that has the potential to harm satellites, equipment, and humans in the near-Earth space environment or on the ground. Warnings apply to phenomena that require a customer to take action in order to protect assets. The Earth-bound analogy is a warning issued for thunderstorms, tornadoes, etc. The key to a warning is the ability to specify when an event will occur, how intense it will be, and how long it will last. Warnings cover the 0- to 24-hour period and are based on observations of causal events (like solar flares), observations of actual events (such as a geomagnetic storm onset), or extrapolations of trends (such as increasing proton fluxes). Most warnings are issued immediately (within minutes) upon observation of a certain event or condition.

A *nowcast* begins with a specification of current conditions, typically based on observations and models that assimilate data and fill the gaps, then projects those conditions into the near future. The time range for this projection is usually slightly shorter than the time required to update the observations. Nowcasts are issued on a regular basis and may include a warning if an event is in progress.

The *forecast* differs from the nowcast in the timeframe it covers and the techniques involved in producing it. Short-term forecasts cover the period from 6 hours to several days. Mid-term forecasts extend these forecasts out to several months. Long-range forecasts can cover some parameters of the solar cycle out through 10 years. Short-term and mid-term forecasts provide general conditions and may also include warning-type events, but they are not intended to provide the timing accuracy necessary to give the fidelity that a customer may want for taking protective action. They are issued on a daily schedule that allows time for the full use of all computing tools, albeit within a rigid timeline. Longer-range solar cycle forecasts are issued monthly and computed automatically.

Post-analysis is used to identify the space weather factors that may have contributed to operational anomalies of systems affected by space weather. Observations are critical for analyzing the state of the environment when the anomaly occurred. Immediate post-analysis is required to identify whether observed anomalous behavior of a system was caused by space weather or by other factors such as mechanical failure, engineering design problems, or software errors. Post-analysis is a valuable tool in providing input into improvements in engineering designs of systems.

Although specification is not a product, it is the starting point for the products described above. Specification refers to the fusion of all available observations into a coherent and realistic representation of the state of the environment at the time of the observations. This step is critical in order to accurately initialize predictive models, perform after-the-fact analyses, and provide the forecaster with a reliable picture of present conditions. The ability to nowcast is only as good as the quality and timeliness of the observational data received and the effectiveness of models in fusing the data into coherent representations of the environment.

2.2 Current Capabilities

This section provides general information on the current operational observing systems and the models supporting forecasting operations at the NOAA Space Environment Center and the Air Force's 55th Space Weather Squadron. Figure 2-1 graphically depicts the current space weather support structure showing both ground and space-based systems and delineating civil, military, and international systems.

The section concludes with an assessment of the overall capabilities of these operational systems. Information on future operational observing systems is contained in Chapter 5 on technology transition. Descriptions of research observing systems and models are in Chapter 3.

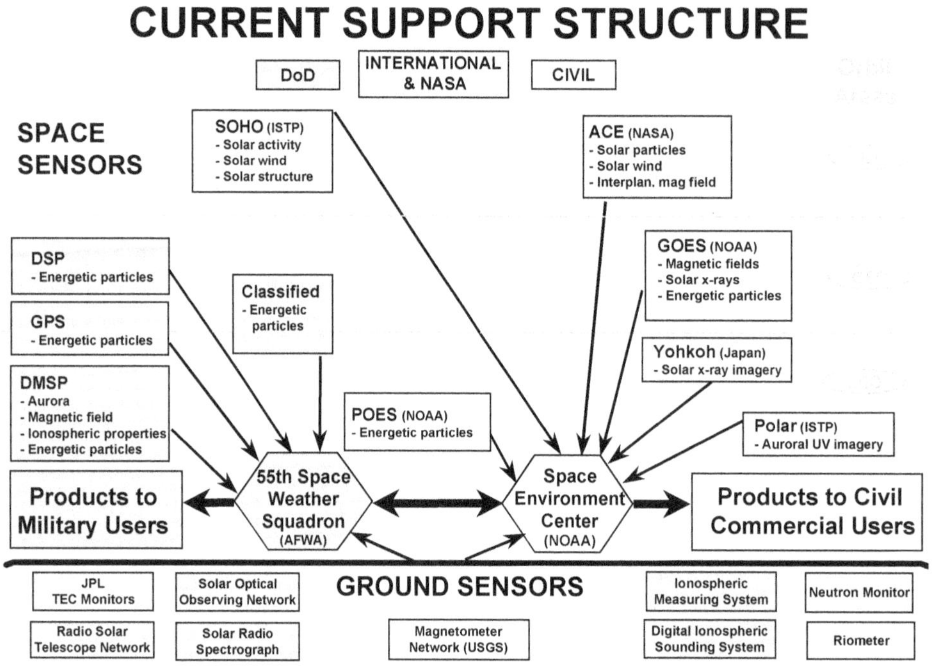

CURRENT SUPPORT STRUCTURE

Figure 2-1. Current Space Weather Support Structure

2.2.1 Operational Observations

GOES Space Environment Monitor. The Space Environment Monitor (SEM) aboard the Geostationary Operational Environmental Satellite (GOES) measures in situ the effect of the Sun on the near-Earth solar-terrestrial electromagnetic environment, providing real-time data to the Space Environment Center (SEC).   The SEM subsystem consists of four instruments used for measurements and monitoring of the near-Earth (geostationary altitude) space environment and for observing the solar x-ray output. An energetic particles sensor (EPS) and high-energy proton and alpha detector (HEPAD) monitor the incident flux density of protons, alpha particles, and electrons over an extensive range of energy levels. Solar output is monitored by an x-ray sensor (XRS) mounted on an x-ray positioning platform, fixed on the solar array yoke. Two redundant three-axis magnetometers, mounted on a deployed 3-meter boom, operate one at a time to monitor Earth's geomagnetic field strength in the vicinity of the spacecraft. The SEM instruments are capable of ground command-selectable, in-flight calibration for monitoring on-orbit performance and ensuring proper operation.

POES Space Environment Monitor. The SEM-2 Space Environment Monitor aboard the NOAA Polar-orbiting Operational Environmental Satellites is a multi-channel charged-particle spectrometer which senses the flux of charged particles at the satellite altitude, and thus contributes to knowledge of the solar-terrestrial environment. SEM-1 units have been in orbit on the TIROS-N series since 1978. SEM-2 consists of two detectors: the total energy detector (TED) and the medium energy proton and electron detector (MEPED) along with a data processing unit (DPU).

DMSP. The Defense Meteorological Satellite Program (DMSP) is now operated by the Department of Commerce as part of the convergence of polar orbiting environmental satellite programs. POES and DMSP will be replaced by the National Polar-orbiting Operational Environmental Satellite System (NPOESS) beginning in about 2008. The DMSP program designs, builds, launches, and maintains several near polar orbiting, Sun synchronous satellites monitoring the meteorological, oceanographic, and solar-terrestrial physics environments. The satellites are Sun synchronous at an altitude of approximately 830 km above the Earth. The orbit period is approximately 101 minutes. Since there are usually two satellites, separated in longitude by approximately six hours of local time, global coverage is provided every six hours.

Each DMSP satellite monitors the atmospheric, oceanographic and solar-geophysical environment of the Earth. The visible and infrared sensors collect images of global cloud distribution across a 3,000 km swath during both daytime and nighttime conditions. The coverage of the microwave imager and sounders is one-half the visible and infrared sensors' coverage, thus they cover the polar regions above 60° latitude on a twice daily basis but the equatorial region on a daily basis. The space environmental sensors record along-track plasma densities, velocities, composition and drifts. The space environment instruments on DMSP are the SSJ/4 (Precipitating Electron and Ion Spectrometer), the SSIES (Ion Scintillation Monitor), and the SSM (Magnetometer).

The data from the DMSP satellites are received and used at operational centers on a continual basis. The data are sent daily to the National Geophysical Data Center (NGDC), Solar Terrestrial Physics Division (STPD) for creation of an archive.

Advanced Composition Explorer (ACE). The Advanced Composition Explorer (ACE) flies at the L1 libration point approximately 1.5 million kilometers sunward from Earth. The ACE RTSW (Real-Time Solar Wind) system provides data to continuously monitor the solar wind and to allow SEC to produce warnings of impending major geomagnetic activity, up to one hour in advance. The RTSW system gathers data at high time resolution from four ACE instruments (MAG - magnetic field vectors, SWEPAM - solar wind ions, EPAM - energetic ions and electrons, and SIS - high-energy particle fluxes), packs the data into a low-rate bit stream, and broadcasts the data continuously. With a combination of dedicated ground stations (Communications Research Laboratory [CRL] in Japan and Rutherford Appleton Laboratory [RAL] in the United Kingdom), and time on existing ground tracking networks (NASA's DSN and the USAF's AFSCN), the RTSW system can receive data 24 hours per day throughout the year. The data are downloaded, processed, and dispersed within five minutes from the time they leave ACE. The raw data are immediately sent from the ground station to the Space Environment Center, processed, and then delivered to the CRL Regional Warning Center at Hiraiso, Japan, to the USAF 55th Space Weather Squadron, and placed on the World Wide Web. The low-energy energetic particles information from ACE RTSW are used to warn of approaching interplanetary shocks and to help monitor the flux of high-energy particles that can produce radiation damage in satellite systems. The ACE data became operational on January 21, 1998, and drive a model developed to predict geomagnetic K-indices up to two hours in advance of their observation at Earth. Alerts and warnings of geomagnetic storms based on ACE data have been implemented in SEC.

Ionosonde Network. The Air Force's 55th Space Weather Squadron operates the USAF Digital Ionospheric Sounding System (DISS) network in order to observe and specify the global ionosphere in real time. There are over a dozen fully automated digital ionosondes deployed worldwide to perform this function. These fully automated ionosondes are derived from the well known Digisondes developed at the University of Massachusetts at Lowell (UML). They are nearly identical to UML's Digisonde 256 (D256), which has been available since the mid-1980s. Additional ionosondes exist to supplement the DISS network, but they do not report observations in real time. Their data satisfy Air Force needs for 6- to 24-hour delayed summaries. The DISS network ionosondes are found primarily in the mid-latitude Northern Hemisphere, particularly in the continental United States.

The DISS network provides data for many USAF products, including specification and forecasts of primary and secondary HF radio propagation characteristics, ionospheric electron density and total electron content, ionospheric scintillation, environmental conditions for spacecraft anomalies, and sunspot number. DISS products are used by the Air Force, Army, Navy, Coast Guard, and the DOD's Unified Commands. NASA Wallops Island and the Arecibo Observatory are also customers for DISS products.

Scintillation Network Decision Aid (SCINDA). SCINDA is a regional nowcasting and short-term forecasting system for UHF and L-band scintillation at equatorial latitudes. It accepts data from ground-based instruments measuring amplitude fluctuations from UHF and L-band receivers and uses these data to generate tailored nowcast and forecast products for tactical users in a graphical depiction. Furthermore, it can accept measurements from DMSP satellites that show areas where scintillation is likely to form in a given night. Basic proof-of-principle validation is currently underway.

Magnetometers. Magnetometers measure variations in the Earth's magnetic field. The United States Geological Survey (USGS) operates one of several networks of ground-based magnetometers. Several of these magnetometers provide the data used at the Air Force's 55th Space Weather Squadron to compute the level of geomagnetic activity in real time. These indices are used to analyze for satellite drag and ionospheric propagation conditions, as well as providing an indication of the strength of currents flowing in the upper atmosphere and the near-Earth space environment.

GPS Receivers. The worldwide network of 170 GPS receivers managed by NASA's Jet Propulsion Laboratory continuously monitors the dual frequency L-band signals from GPS satellites. Each receiver is capable of receiving signals from more than eight satellites simultaneously. The data collected from this network include carrier phase and group delay to derive the total electron content (TEC) of the ionosphere along these different ray paths to the satellites. The TEC data are interpolated to obtain Global

Ionospheric Maps (GIM) of TEC every 15 minutes. Differential maps indicating the percentage change of TEC between magnetically quiet and disturbed periods have also been generated to monitor storm-induced TEC perturbations every 15 minutes.

Ground-based Coronagraphs. Coronagraphs provide images of the Sun's corona, the outermost layer of the solar atmosphere. Ground-based coronagraphs complement space-based instruments. A white light coronometer is located at the Mauna Loa Solar Observatory in Hawaii, operated by the High Altitude Observatory, National Center for Atmospheric Research. A hydrogen-alpha coronograph is located at the Pic du Midi Observatory in France. Coronagraphs provide observations of coronal mass ejections.

Ground-based solar optical and radio observatories. The Air Force's Solar Electro-Optical Network (SEON) consists of both solar radio and optical telescopes. The radio telescopes provide information on the level of solar activity by monitoring radio noise. Intensities of solar radio emission during quiet and flare times are measured over the frequency range of 30 kHz to 100 GHz. Interpretation of the data is done on-site and forwarded to the forecast center in terms of Type I to Type V radio flare emission and in terms of impulsive (seconds to 10 min) or gradual (10 mins to days) events. The optical telescopes are used to make images in H-alpha and white light, as well as line-of-sight magnetograms of the surface magnetic field. These data are analyzed on-site and messages are forwarded to the forecast center, giving the current level of flare activity and filament disappearances that can lead to coronal mass ejections. Depending on the complexity of phenomena observed for the active regions, a 24-hour flare warning is issued.

2.2.2   Operational Models at the Space Environment Center

Magnetospheric Specification Model. The magnetospheric specification model developed for operational use by Rice University, with funding from the USAF and start-up funding from NOAA, is run on a regular production basis using geomagnetic indices as driver data. The model currently runs every three hours as an operational tool to provide retrospective and real-time maps of charged particle fluxes throughout the inner and middle magnetosphere. Input data from satellite and ground stations drive the model, whose primary input parameters may include Kp, magnetopause standoff distance, polar-cap potential drop, auroral boundary index, and Dst. In addition, the model can operate with reduced suites of input data and will run, if necessary, from Kp alone. The model follows particle drifts through the magnetosphere using time-dependent electric and magnetic field models while keeping track of loss by charge exchange and electron precipitation into the ionosphere. It assumes particle transport by $\mathbf{E} \times \mathbf{B}$, gradient, and curvature drift. An isotropic particle distribution is assumed to be maintained by pitch-angle scattering mechanisms that do not change particle energy. Data-based algorithms are used to specify initial condition and boundary condition particle fluxes. The model is designed to specify fluxes of electrons in the energy range responsible for spacecraft charging, ~100 eV to ~100 keV.

PROTONS. PROTONS is used to predict the temporal parameters and intensity of solar proton events once activity is observed on the Sun. The model is driven by soft-ray observations from the GOES 1-8 nanometer Space Environment Monitor and by observations from the ground-based Solar Electro-Optical Network including radio burst type (Type II and Type IV) and solar flare position. It produces a prediction of the delay and rise time to maximum and the maximum intensity of the flux of protons with energy greater than 10 MeV. The model is based on the use of parametric solutions to particle propagation equations, which are fit to actual observations of solar and energetic particle observations.

Wang and Sheeley Model. The Wang/Sheeley model is used in the SEC to predict the background solar wind speed and IMF polarity at Earth 3 to 4 days in advance. This forecasting tool is a modified version of the original Wang/Sheeley model. Using a traditional approach, the Wang/Sheeley model makes daily predictions at 1 AU for the next solar rotation based upon synoptic maps of the photospheric field from the previous rotation. Arge/Pizzo have modified the Wang/Sheeley model used in the SEC so as to take full advantage of the most recent photospheric magnetic data available; their implementation updates the synoptic charts with new magnetograms as frequently as possible and makes new predictions after each update. This modified model uses an empirically deduced function to relate the solar wind speed at the source surface with the coronal field expansion factor derived from a source surface model. The coronal divergence factors are determined by tracing the magnetic field lines down to the photosphere and calculating the ratio of the magnetic field strengths at the endpoints of each magnetic field line, relative to what is expected for purely radial expansion. The solar wind is then propagated from the source surface out to Earth using a simple mass-flux conservative algorithm to account for stream interactions along the way.

Costello Model For Predicting Kp. As implemented in SEC, the model uses ACE real time solar wind data to predict Kp. The current prediction algorithm is the Costello Neural Network (CNN) [Kirt Costello, PhD Thesis, Rice University, 1997]. Input to the CNN for each Kp output consists of two consecutive 1-hour averages of three solar wind parameters, Velocity (V), IMF magnitude (B-total), and IMF Bz component (Bz, in GSM coordinates). Effectively, pKp is computed at the location of ACE and the prediction is propagated to Earth by adding a computed lead-time to the current time. The lead-time is computed by dividing ACE's upstream distance (X-coordinate) by the measured plasma velocity, V. Lead-times are usually in the range of 30 to 60 minutes.

2.2.3   Operational Models at the 55ᵗʰ Space Weather Squadron

Ionospheric Activity Index (IACTIN). Provides ionospheric corrections for regional areas.

Real-Time Ionospheric Correction Maps from a CTIM. A coupled thermosphere ionosphere model (CTIM) is being used to produce an ionospheric correction map for the peak F-region electron density (NmF2) and total electron content (TEC), at middle to high latitudes. The correction map is designed to scale climatological values of NmF2 and TEC, for a given solar flux and season, in order to correct for the effects of geomagnetic disturbances including severe storms. Geomagnetic activity can cause large regional depletions (negative phase) or increases (positive phase) of electron concentration, with a distribution that depends on season and local time. The model is driven by the real-time auroral power index (PI) derived from measurement by the TIROS/NOAA polar orbiting satellite. The PI is an estimate of the auroral power input to one hemisphere of the Earth and is updated about every orbit (1.5 hours). PI values are calculated from the satellite data with a time resolution of about 45 minutes and are then used to drive the physical model to update the ionospheric correction maps.

Bent Model. This ionospheric model performs ray traces through detailed electron density profiles and is used to calculate TEC for radar signal ionospheric corrections at 55th SWXS. This analysis allows space operators to accurately locate satellite objects and to conduct threat detection propagation. The Bent Model accurately describes the ionosphere to obtain high precision radar signal delay and directional change due to refraction. Tile data upon which the Bent Model is based was collected from 1962 to 1969 and includes the solar cycle maximum and minimum. Bent Model inputs include the date, Universal Time, transmitter/receiver locations, operating frequency, space vehicle elevation and altitude rate of change, solar flux, and sunspot number. Model outputs include vertical and total electron content above the transmitter, a vertical electron density profile, and the total electron content along the path between the satellite and the tracking site.

Improved Auroral Prediction Model (IAPM). This model provides the capability to continually monitor the global position of the auroral precipitation boundary and the position of the boundary relative to Air Force installations or systems of interest. The program is a command driven system which builds and plots the auroral climatology relative to the surface of the Earth. It provides the capability to continually monitor the global position of an auroral precipitation boundary and the position of the boundary relative to Air Force installations or systems of interest. Plots of the aurora and its boundaries are based on Defense Meteorological Satellite Program (DMSP) electrostatic sensor SSJ/4 data or geomagnetic index Kp input to an electron precipitation model. The program is interactive, started manually by the user and controlled by commands or series of commands stored as macros. Options provide the capability for the user to define grid points, triangles, plotting grids, and various parameters such as date and time.

Ionospheric Communications Enhanced Profile Analysis and Circuit (ICEPAC). ICEPAC predicts the expected performance of high frequency (HF) broadcast systems, and in doing so is useful in the planning and operation of HF transmissions for the four seasons, different sunspot activities, hours of the day, and geographic location. The ICEPAC computer program is an integrated system of subroutines designed to predict high-frequency (HF) sky-wave system performance and analyze ionospheric parameters.

Ionospheric Communication Analysis and Prediction Program (IONCAP). This program predicts the maximum useable communication frequency (MUF), the frequency of optimum transmission (FOT), and the lowest useable frequency (LUF) between two transmit and receive points. It is one of several climatological ionospheric models used for applications to generate tailored warfighter products. In the IONCAP model, the numerical coefficients are functions of geographic latitude for both solar maximum and minimum. It also takes into account the retardation below the F2 layer. For the MUF computations the model uses the corrected form of Martyn's theorem. As the absorption equations using the secant law do not work for lower frequencies at altitudes below 90 km, these equations have been modified in the IONCAP program. The IONCAP provides two programs: 1) the ITS-78 (see more information below in the ITS-78 section) on short path geometry, and 2) the path >10,000 km geometry. In addition to the ITS-78 model, the path computations now include the F1 mode, the over-the-MUF mode, D and E region absorption losses, and sporadic E losses. A correction to frequency dependence is added for low frequencies reflected from altitudes below 90km.

Magnetospheric Specification Model (MSM). See description in Section 2.2.2.

Magnetospheric Specification and Forecasting Model (MSFM). The MSFM provides a nowcast of magnetospheric conditions based on first principles of physics. It is intended to provide accurate information to help prevent and diagnose spacecraft malfunctions due to charging and other environmental causes in real time. It is also capable of forecasting low-energy electron fluxes at geosynchronous altitudes for space weather forecasting.

The MSFM forecasts the near-Earth magnetospheric conditions. Information provided focuses on the fluxes of inner magnetospheric electrons, protons, and O+ ions in the energy range of up to 100KeV. This information is of value to organizations having satellites in orbit about the Earth. The AFRL space weather forecasting codes tell the operators when a satellite might encounter problems, such as disruption to electronic and communication components, due to surface charges caused by the space environment. The value of the model is that it can "read the environment" to determine if the space environmental conditions may be conducive to temporary disruption or malfunction of satellites. Once the source of the problem has been identified, operators can then circumvent the problem by finding an alternate means of collecting data from the satellite.

Parameterized Real-Time Ionospheric Specification Model (PRISM). The ultimate goal of PRISM is to provide increased accuracy for current electron density profiles, atmospheric densities, and auroral disturbances. Operators of HF and satellite communications, radar navigation systems, and operators concerned with predicting satellite orbits already use current and predicted space weather information to more efficiently operate their assets. Daily, worldwide mapping of large volumes of refreshed ionospheric data can reduce modeling errors significantly. Inputs of real-time ionospheric data can potentially improve PRISM output data. The Parameterized Real-Time Ionospheric Specification Model (PRISM) was developed by Computational Physics Incorporated (CPI) for use by the Air Force 55th Space Weather Squadron. PRISM's purpose is to provide an accurate real-time ionospheric specification for DOD use. "Ionospheric specification" specifies the state of the ionosphere globally or regionally at a given time in terms of electron density profile parameters, actual electron density profiles, or both. The most common application of PRISM is through the use of real-time measurements of certain ionospheric parameters to update a database of electron density profiles. The database is generated using specialized software that models various regions of the ionosphere based on physical principles. This database can be updated with data received from a number of different sensors at any time and the effects of the ionosphere on tracking measurements are computed by integrating the total electron current (TEC) along the appropriate line-of-sight.

WideBand Model (WBMOD). WBMOD is a radio frequency ionospheric scintillation code which specifies scintillation parameters, as a function of a variety of geophysical parameters, between any location on the globe and a satellite above 100 km altitude for any frequency above 100 MHz. This information is used to support military systems for communications, command and control, navigation, and surveillance that depend on reliable and relatively noise-free transmission of radio wave signals through Earth's ionosphere. Small-scale irregularities in the ionosphere's density can cause severe distortion of both the amplitude and phase of these signals. WBMOD is used to assess and forecast the radio wave scintillation resulting from these irregularities. The model was developed based on analysis of data from the Defense Nuclear Agency (DNA) Wideband satellite experiment. Due to the limited coverage of the data used in developing the model (a single station at high latitudes and two stations at equatorial latitudes), a recent validation of the WBMOD model showed that it was deficient in a number of areas.

Proton Prediction System (PPS). PPS has been constructed to model protons that are accelerated during energetic solar events (i.e., those associated with flares) near the solar surface. It does not model proton acceleration that might occur in an interplanetary shock propagating towards Earth. The model generates a computerized time-intensity profile of the solar proton intensity expected at the Earth after the occurrence of a significant solar flare on the Sun. Predictions of solar protons, alpha particles, and iron nuclei are available. This forecast is based on location of the solar flare in the heliocentric coordinates and the magnitude of the Sun's output in x-ray or radio wavelengths. Data collected from numerous spacecraft formulate the basis for the mathematical expression relating the Sun's output to the flux of energetic solar particles during the course of an event. Inputs to the model are solar flare time, location, and intensity. The flare intensity may be input using x-ray or radio data. The outputs of PPS include the polar cap riometer absorption maximum and time, high altitude radiation dose level and time, Extra-Vehicular Activity (EVA) radiation dose level and time, protons (greater than 10 p/cc greater than 5,10, and 50 MeV) maximum flux and time, and total fluence.

Ramsey-Bussey Total Electron Content (RBTEC). The Ramsey-Bussey Total Electron Content [RBTEC (APBA)] program was developed as an effort to produce a realistic electron density profile (EDP) based on parameters which can be forecast reasonably accurately. The output produced by the APBA program is used to aid in predicting errors occurring in range and azimuth of satellite tracking radar due to the effects of ionospheric retardation and refraction. The model was developed to predict electron densities from 0 to 1000 km altitude.

Shock Time of Arrival (STOA). This model forecasts the time of arrival of a solar wind shock at the Earth based on solar x-ray and radio data. The STOA model is a physically-based model derived from three-dimensional magnetohydrodynamic simulations. The model inputs are flare time, location, type II radio burst speed, solar wind background velocity, and flare intensity in x-rays and decay times. Model outputs include the time of arrival at the observer (default is Earth - 1 AU), the speed of the shock (Mach number), the total propagation time, and a two-dimensional shock front map at 3-hour intervals.

Global Ionospheric Forecast Model (IFM). The Air Force's three-dimensional, time-dependent model of the global ionosphere was streamlined in an effort to develop a computationally fast, user friendly, reliable, Ionospheric Forecast Model (IFM). The model yields predictions for the molecular and oxygen ion densities and the ion and electron temperatures over the globe at E and F region altitudes. The model also contains a simple algorithm for predicting H+ densities in the F region. The inputs needed by the IFM are global distributions of the neutral densities, temperatures, and winds, the auroral oval precipitation, the magnetospheric and dynamo electric fields, and the topside electron heat flux. Because of the model's modular construction, the IFM can readily accept different global input patterns and, hence, has the capability of being driven by real-time inputs from Air Force satellites or ground-based sites. In the current version of the model, the input patterns have been selected and the IFM is therefore self-contained and can be driven by simple geophysical indices.

Combined Release and Radiation Effects Satellite Electron Model (CRRESELE). CRRESELE is used to map the electron flux models into a three-dimensional grid specified by the user. CRRESELE predicts the omni-directional electron fluences for 10 energy intervals (0.5-6.6 MeV). The CRRESELE software uses flux models created from data collected by the high-energy electron fluxmeter on board the CRRES system. CRRESELE is actually eight different electron fluence models. Six of the eight are parameterized by geomagnetic data. The other two models are for average conditions and a model for the maximum flux values. CRRESELE inputs include magnetic field model choice, energy channel (10 levels with central energies between 0.65 - 5.75 MeV electrons), and geomagnetic activity level (Ap, AVE, or MAX). The CRRESELE output is a 3-D gridded data set of proton flux for the selected energy channel and activity level (including AVE and MAX) in units of #/cm2/sec/KeV. The CRRESELE model is accessible as a science model in GEOSpace, a workstation software suite developed by the Air Force's Phillips Laboratory.

Combined Release and Radiation Effect Satellite Proton Flux Model (CRRESPRO). CRRESPRO is used to map the proton flux data sets used to create a three-dimensional grid specified by the users. CRRESPRO science and application modules are a UNIX port of the PC program developed and released by the Air Force Research Laboratory. CRRESPRO predicts the proton omni-directional fluence per year and integral omni-directional fluence per year at selected energies in the range 1-100 MeV for an orbit that the user specifies. The CRRESPRO software uses flux models created from data collected by the proton telescope on board the CRRES system. CRRESPRO inputs include magnetic field model choice, energy channel (22 levels between 1-100 MeV protons), and active or quiet conditions. Active conditions use data after the March 1991 storm that produced a significant third radiation belt. Quiet conditions use the data before the storm. The CRRESPRO output is a 3-D gridded data set of proton flux for the selected energy channel and activity level in units of #/cm2/sec/MeV. The CRRESPRO model is also accessible as a science model in the GEOSpace suite.

Combined Release and Radiation Effects Satellite Radiation Dose Model (CRRESRAD). CRRESRAD measures the energy deposited by radiation in a material per unit mass of the material. CRRESRAD predicts the amount of radiation received in a specific orbit by various levels of aluminum shielding during a user-specified time period. The prediction is based on empirical models of accumulated radiation measured by the Space Radiation Dosimeter on board the CRRES. The minimum energies required for particles to penetrate the domes of various thicknesses and accumulate doses in the silicon detectors underneath were 20, 35, 50 and 75 MeV for protons and 1, 2.5, 5 and 10 MeV for electrons. Geomagnetic activity is classified as quiet (before the March 1991 storm), active (after the March 1991 storm), or average (average of quiet and active data). CRRESRAD inputs include magnetic field model choice, hemisphere, channel (LOLET or HILET), and geomagnetic activity (QUIET, ACTIVE, or AVERAGE). The CRRESRAD output is a 3-D gridded data set of the dose rate in units of Rads/Si/sec for the selected shielding level and channel. The CRRESRAD model is also accessible as a science model in the GEOSpace suite.

International Union of Radio Science Coefficients for CY88 (URSI-88). The establishment of a new set of coefficients to numerically map the global variation of foF2 has been developed by a working group under the auspices of the International Union of Radio Science. Global maps that represent the monthly median behavior of ionospheric parameters have been an integral part of ionospheric modeling efforts and propagation prediction methods for over 25 years. Following the 1966 work of Jones and Gallet to adopt a set of CCIR numerical coefficients that could be used to represent the monthly median critical frequency of the F2 region, foF2 can be calculated at any point on the globe at a given universal time. These coefficients were determined from a spherical harmonic analysis of data observed between 1954 and 1958 at over 150 ionosonde locations around the world. In 1970, the CCIR adopted yet another set of coefficients derived from a similar data set but containing a better representation of the solar cycle variation of foF2.

ITS-78 Ionospheric Model. This is the most widely used numerical model for ionospheric predictions. ITS-78 is a major climatological ionospheric model used to generate circuit operational parameters such as the maximum usable frequency (MUF), optimum traffic frequency (FOT), and the lowest usable frequency (LUF). The ITS-78 model and its computer programs were developed by the Institute of Telecommunication Sciences, Earth Sciences Services Administration, Boulder, Colorado. The model is based on the presentation of the ionospheric characteristics in a form of synoptic numerical coefficients developed by Jones and Gallet (1960) and improved by Jones. The important features of the ITS-78 model are the parameters for the D, E, sporadic E and F2 layers of the ionosphere.

Orbit-Application Module (ORBIT-APP). The Orbit application module in GEOSpace provides an interface to the orbit generation and prediction codes. The interface and code generation are shared among several modules requiring code generation. The ORBIT application module provides an interface to the LOKANGL and SGP4 or bit generation and prediction codes. Other modules using the orbit generation modules are the CRRESRAD, CRRESPRO, and CRRESELE applications. Input to the ORBIT application comprises orbital elements and start and stop times of the orbit interval to be predicted. Several methods for inputting the orbital elements are available. The input window is logically divided into three areas: a propagator/element type section, an orbital element input section, and an auxiliary input area. The ORBIT application allows the orbit to be generated by using either the LOKANGLE or SGP4 orbit propagator codes. In addition, the orbital elements to be used by the propagator may be specified in a variety of ways.

Magnetic Field Models (MFM). The Earth's magnetic field is usually modeled as the sum of the "main" internal magnetic field, which is, for the most part, attributed to currents within Earth's core, and an "external" field, attributed to ionospheric and magnetospheric currents. The magnetic field of the Earth dominates over the Interplanetary Magnetic Field (IMF) in the near-Earth environment area called the magnetosphere. Several magnetic field models have been developed to specify the field in the magnetosphere. The code calculates corrected geomagnetic (CGM) coordinates and several other geomagnetic field parameters for geographically specified points on Earth's surface or in near-Earth space. The underlying geomagnetic field is the Definite/International Geomagnetic Reference Field (DRG/IGRF).

Parameterized Ionospheric Specification Model (PIM). PIM is the base ionospheric model on which the Parameterized Real-time Specification Model (PRISM) operates. PRISM uses data from ground-based and satellite-based sources to adjust the parameterized model, giving a near real-time specification of the ionosphere. The PIM science module is a relatively fast global ionospheric model based on the combined output of several physical ionospheric models.

2.3 Assessment of Current Capabilities

Table 2-1 illustrates current capabilities in each space weather region to warn, nowcast, forecast, and provide post-analysis products for space weather events. Red means there is no capability to meet the requirements for the events in the given region, yellow/red indicates very limited capability, and yellow indicates some capability short of meeting operational requirements. No areas are coded green because user-specified needs cannot be met in any area at the present time.

Table 2-1. Current Capabilities Based on Requirements

	Warning	Nowcast	Forecast	Post-Analysis
Solar/Interplanetary	Yellow/Red	Yellow/Red	Yellow/Red	Yellow
Magnetosphere	Red	Yellow/Red	Red	Yellow/Red
Ionosphere	Red	Yellow/Red	Red	Yellow
Neutral Atmosphere	Red	Yellow/Red	Red	Yellow/Red

The following paragraphs summarize the current capabilities in each of the four product areas.

Warnings. Very little capability exists to warn for space weather events. Causal solar events can be detected in real time, but warnings based on these events lack sufficient reliability for immediate mitigation actions and do not provide useful lead time or information on magnitude and duration of the event. Capability is strongest (albeit very limited) in the solar/interplanetary region because of the 24-hour observing system of solar observatories.

Nowcasts. Limited nowcasting capability based on rudimentary models exists at operational centers. However, the models offer little capability beyond information available from empirical methods and climatology. Capability is best when data to initialize the models are received in a timely manner.

Forecasts. Forecasting capability suffers from the same weaknesses as warning capability, and in addition the challenge is greater because forecasting requires longer lead times. This in turn requires a more complete understanding of both the solar events that drive space weather and the way the space environment reacts to those events.

Post-Analyses. Current capabilities are the strongest in support of post-analysis requirements; however, significant deficiencies still exist. The relatively strong capability in this area derives from the fact that some post-analyses are not required in real time. This allows the analyst to gather data that may not have been immediately available to operators and to assimilate it at leisure.

In summary, these limited capabilities come from a basic understanding of space weather combined with a limited observation base and still rudimentary computer models. However, they lack the necessary accuracy and four-dimensional detail to meet operational requirements.

2.4 Assessing Capabilities with Metrics

A "space weather metric" is a quantitative measure of the ability of a scientific algorithm or model to predict or nowcast the value of a physical parameter involved in space weather. A specific metric has three elements:

- A parameter defined at some position and time (for example, the F-region peak electron density at mid-latitude every hour for the next day).
- An observable to which a prediction can be compared (e.g., density measurement by an incoherent-backscatter radar facility).
- A criterion by which the metric is quantified (e.g., RMS difference between prediction and observation).

The Space Environment Center (SEC) of the National Oceanic and Atmospheric Administration (NOAA) and the U. S. Air Force 55th Space Weather Squadron provide space forecasts and nowcasts for a wide variety of practical applications. Those agencies have developed "application metrics" for measuring the value and validity of the specific services they provide. However, for the purpose of measuring the overall progress of the NSWP, it is useful for the scientific community to define a separate and broader set of metrics, for the following reasons:

- Application metrics change as technologies change. Metrics for the NSWP must remain valid at least for the ten-year life span of the Program.
- Scientific metrics must be open to the scientific community. Although application metrics developed by the NOAA Space Environment Center are public and involve no proprietary information, some applications metrics often involve defense secrets (military) or trade secrets (commercial users).
- Although there is remarkable overlap between parameters that are important to the application community and scientifically important parameters, the overlap is not 100%. For example, ring-current ions are an important element of magnetospheric physics but have little direct effect on present technological systems.
- To measure progress, scientific metrics should have a scale that encompasses both presently available scientific algorithms and the best that we could hope for by the end of the NSWP. Present algorithms are not good enough to make useful predictions of all aspects of space weather and might thus score zero on some application metrics. There is also a chance that our ability to predict some parameters will, by the end of the NSWP, exceed what is needed for present technologies. A good scientific metric should encompass both extremes.

Although metrics play a major role in some fields of physical science (notably meteorology), their use is far from universal and most space scientists are unused to dealing with them. To study the use of metrics for assessing progress in the NSWP, the National Science Foundation convened three study panels. The metrics study panels were designed to include representatives from NOAA/SEC and the Air Force who are familiar with the needs of the users of space weather services, and to include some people with experience with metrics. A major goal of the study was to acquaint the space physics research community with the idea of metrics and to initiate scientific discussion of the subject. To this end, a special session was held at the Fall 1997 meeting of the American Geophysical Union, and metrics presentations have been given at three other large meetings.

The panels recommended that the following types of metric evaluations be undertaken as soon as possible to establish a regular program of scientific metrics for space weather capabilities:

Type 1.   Measurements should be made at a regular cadence and compared systematically with algorithm or model predictions to establish statistically valid baseline metrics. In addition to the overall averages, average errors should be recorded for different conditions (e.g., *Kp* levels in the case of magnetosphere-ionosphere metrics).

Type 2.   When groups of scientists carry out event studies, comparing various models and other algorithms to observations, they should evaluate their models and algorithms in terms of the same standard metrics used in the statistical analyses of Type 1. This would help to tie event studies and campaigns, which are a regular feature of cooperative research programs (e.g. CEDAR, GEM, SHINE) as well as some NASA spacecraft programs, to progress of the NSWP.

Care must be taken to withhold from the input suite any data that will be used to test the model. In some cases, withholding those data may artificially degrade model performance. This is an advantage of testing of Type 2 as part of scientific campaigns, which generally utilize more data than is routinely available at the forecast centers. Those additional data could be used for testing without degradation of the input stream.

This metrics study was divided into three disciplinary areas: Ionosphere-Thermosphere (I-T), Magnetosphere-Ionosphere (M-I), and Solar-Interplanetary (S-I). The following paragraphs list metrics appropriate for each area as well as concerns specific to each. The physical systems are, of course, closely coupled. A physical model of Earth's magnetosphere must be driven by information about the solar wind, and a physical model of the Earth's ionosphere and thermosphere must be driven by inputs from both the magnetosphere and the Sun. Therefore, some of the metrics listed in the M-I section were, in fact, forced by the needs of I-T models, and all of the metrics listed in the S-I section represent requirements of the M-I and I-T models.

2.4.1   Ionosphere-Thermosphere Metrics

The principal ionosphere-thermosphere parameters that need to be predicted are shown in Table 2-2, grouped according to priority. Ability to forecast or nowcast ionosphere-thermosphere weather should be judged according to ability to deal with the parameters listed.

For each physical parameter, metrics should be defined that measure ability to forecast and nowcast the climatological mean, the one-sigma limits in the daily values ("day to day variability"), a particular time interval (e.g., one-day forecast), and the departure from the climatological mean over a particular interval. An additional set of metrics is needed to specify and forecast macroscopic features that can dominate certain regions of the ionosphere-thermosphere domain, including the Appleton anomaly, high-latitude features (sub-auroral trough, tongues and holes in polar cap ionization, neutral density holes), equatorial pre-reversal enhancement in vertical ion drift, transient ionospheric disturbances (TIDs), and the ratio of atomic oxygen to molecular nitrogen column abundance.

Table 2-2. Priority List of Key Physical Parameters for the Ionosphere and Thermosphere

First Priority:
Electron density N_e, including intrinsic variability
Neutral mass density ρ, including intrinsic variability
$\delta N_e / N_e$, the amplitude of the electron density irregularities
Second Priority:
Neutral and ion composition
Thermospheric winds and temperatures
Low-latitude ion drifts
Third Priority:
Electron and ion temperature
Fourth Priority:
Minor species

No single metric adequately represents our overall ability to forecast and nowcast the state of the ionosphere and thermosphere. Table 2-3 presents a focused set of five. The best single metric is judged to be the first entry of Table 2-3: the RMS error in the electron density from 200 km to 600 km, for stations at low, middle, and high latitudes. The selection in Table 2-3 is based on two criteria – the importance of the parameter in describing the state and condition of the I-T system, and the availability of routine, accurate measurements for quantifying our forecast and nowcast capabilities. The details for the first high priority metric from the table above are described in Figure 2-2.

Table 2-3. Priority Ionosphere-Thermosphere Metrics

Category	Parameter	Place	Time	Cadence	Data	Criterion
F-region ionosphere	*Minimum:* NmF2, hmF2 *Desired:* N$_e$(200-600), Δh ~20 km	Low, mid, and high latitudes	03-09 LT 09-15 LT 15-21 LT 21-03 LT	Hourly	Ionosonde or incoherent scatter radar (Jicamarca, Arecibo, Millstone Hill, and Sondre Stromfjord)	RMSE
High-latitude structure	N$_e$ (~800 km)	Orbit plane of polar satellite Poleward of 45 deg mag lat Δx ~100 km	Every orbit	Every orbit	DMSP - SSIES	RMSE
Pre-reversal enhancement	Peak magnitude of vertical ion drift V$_i$ (400 km)	Magnetic equator	16-20 LT	Daily	Incoherent scatter radar (Jicamarca)	Obs-model or RMSE
Scintillation/ Ionospheric irregularities	σ_ϕ, S$_4$ at 250 MHz and 1 GHz	Between +20 and -20 deg dip latitude	18-04 LT Δt ~ 1 hr	Daily	Geostationary and GPS satellites	RMSE
Electron content	Peak TEC and N/S latitude location of Equatorial Ionization Anomaly	N/A	Every orbit of observation	Every orbit of observation	TOPEX	Obs-model/ obs or RMSE

The four incoherent scatter radars of the U. S. meridional chain represent a convenient means to routinely measure electron densities in the altitude range 200 km to 600 km with 20 km altitude resolution. Because the radars are operated for at least 24 hours continuously approximately each month in support of the World Day experiments, metrics can be determined on nearly a monthly basis. This ensures adequate sampling as a function of season. The numerous observation periods per year will also ensure good sampling during quiet, moderate, and disturbed geomagnetic periods.

The NSWP ionosphere-thermosphere metric to be determined for each of the four incoherent scatter radars is given by the following formula:

$$\Delta = \frac{1}{24 \times 21} \sum_{t=0}^{23} \left\{ \sum_{h=200,20}^{600} [n_{o;t}(h) - n_{m;t}(h)]^2 \right\}^{1/2}$$

Where $n_{o;t}(h)$ is the observed density at altitude h and time t and $n_{m;t}(h)$ is the corresponding model value. This gives the RMS error in the density measured/computed hourly at 21 altitudes, averaged over 24 hours.

Allowed inputs for the models include A_p, K_p, F10.7, and all normally available satellite data.

Figure 2-2. NSWP Ionosphere-Thermosphere Metric 1

2.4.2   Magnetosphere-Ionosphere Metrics

Table 2-4 lists major aspects of the Earth's magnetosphere and its coupling to the ionosphere. A central difficult issue for the Magnetosphere-Ionosphere Panel has been how to cover all of the important aspects of the many-faceted system with just a few metrics. The aspects listed are coupled to each other in various ways, but the ability to forecast one does not imply the ability to forecast others. The dynamical behavior of the radiation belts, for example, is quite different from the dynamical behavior of the aurora or the plasmasphere.

Table 2-5 lists top-priority magnetospheric metrics as well as details related to data collection methods. The ionospheric electric field and the precipitating electron fluxes were chosen partly because of their importance as inputs for ionosphere-thermosphere models. Several magnetic indices were included because of their wide use and because they are designed to indicate global conditions. Magnetospheric electron fluxes were

Table 2-4. Major Features of the Magnetosphere-Ionosphere Coupled System

Feature	Includes
Magnetic field configuration	Global magnetic structure, including dayside, tail; ground magnetic variations
Electric field configuration	Ionospheric and magnetospheric. Represents effects of solar-wind/magnetosphere coupling, magnetospheric convection
Auroral precipitation	Precipitation from polar cusp, polar cap, main auroral zones and plasma sheet
Trapped energetic particles	Includes ring current and inner and outer radiation belts, from ~ 1 keV to ~ 100 MeV
Cold particles	Plasmasphere, plasmapause, suprathermal ions
Plasma sheet, plasma-sheet boundary layer	Kilovolt electrons and ions that extend into the tail
Magnetopause	Shape and position, reconnection, transfer processes, boundary layers
Waves and small-scale effects	Cause particle loss by pitch-angle scattering, allow magnetic reconnection, accelerate auroral particles

included because of their importance as space weather parameters. The list is restricted to the top-priority parameters that are regularly measured by ground stations or full-time monitoring spacecraft. Since the relevant observing stations, whether space- or ground-based, operate continuously, the comparisons should be made continuously.

The best single metric for the magnetosphere-ionosphere system is the ionospheric electric field, as specified in the top line of Table 2-5, and described in Figure 2-3. This metric covers nearly all magnetospheric field lines and a wide range of physical processes. It combines magnetospheric convection, substorm effects, magnetic storms, polar cap phenomena, and, to a modest extent, low-latitude effects. Convection has a major effect on the cold plasma structure and dominates ring-current injection. However, the metric defined in the top row of Table 2-5 is certainly not comprehensive. For example, it is not an indicator of magnetopause position or of the state of the radiation belts. The ionospheric electric field is predicted by a range of empirical and first-principles models.

Table 2-5. Priority Magnetosphere-Ionosphere Metrics

Category	Parameter(s)	Place	Averaging interval	Data	Criterion*
High-latitude ionospheric electric field	Component of **E** along track of polar-orbiting spacecraft above 50 deg invariant latitude	~ 1000 km altitude, from dawn-dusk orbit	100 km along s/c track	Ion drift meter on DSMP spacecraft	Mean absolute error in component of **E** along satellite path
Auroral electron flux	Latitude-integrated energy flux, number flux. Latitudinal centroid of energy flux	~ 1000 km altitude, from nightside auroral zone crossings.	100 km along s/c track	Precipitating electron flux measured by DMSP or NOAA spacecraft	Mean absolute error
Magnetic indices	AE (electrojets) Dst (ring current) Kp (overall activity)	Ground stations	Time resolution of index	Ground magnetometers	Mean absolute error
Magnetospheric electron fluxes	Fluxes of > 10 keV and > 1 MeV electrons	Geosynchronous orbit	15 minutes	LANL and NOAA spacecraft	Mean absolute error in log(flux)

*Mean absolute error $= <|F_{predicted} - F_{observed}|>$

Let $v_{\perp o}$ be the observed drift velocity perpendicular to both the satellite trajectory and the magnetic field. Divide the satellite track into 100 km segments. In each segment compute the average value of $v_{\perp o}$.

Let $\mathbf{E}_m$ be the electric field vector from a model. Then

$$v_{\perp m} = \frac{\mathbf{E}_m \cdot \hat{\mathbf{s}}}{|\mathbf{B}|\sin\psi}$$

where $\hat{\mathbf{s}}$ is the unit vector pointing along the satellite track and ψ is the angle between the magnetic field and the satellite trajectory.

Average the model velocity in the same segments as were used for the observed velocities. Calculate the value of the metric for each satellite pass by:

$$\Delta = \frac{1}{N}\sum_{i=1}^{N}\left|v_{\perp o;i} - v_{\perp m;i}\right|$$

If desired, the per pass metric can be averaged over multiple satellite passes to derive a single metric for an event.

The allowed inputs include the solar wind velocity, density, and IMF at the L1 libration point as well as Kp and Dst. These models may **not** use ground magnetometer measurements, electric field or plasma drift measurements.

If the model uses solar wind inputs, it must have a well-defined algorithm for determining the appropriate delay between the L1 measurement and the ionospheric response. In particular, the model may **not** determine the delay by finding the delay that gives the best comparison between model and observation.

Figure 2-3. NSWP Magnetospheric Metric 1

2.4.3　Solar-Interplanetary Metrics

Crucial solar and solar-wind parameters that are required by physical models of the thermosphere, ionosphere, magnetosphere are listed in Table 2-6, along with proposed metrics. The primary metric is described in Figure 2-4.

Table 2-6. Priority Solar and Interplanetary Metrics

Category	Parameters	Observing Location	Averaging Interval	Data Source	Criterion	Needed for
Solar EUV	Intensity of strong spectral lines (e.g., 30.4 nm) Integrated EUV flux	L1	1 day	SOHO	RMSE	Ionosphere and thermosphere models
Solar x-rays	Intensity of 0.1-0.8 nm flux	Earth orbit	1 hour	GOES	RMSE	Ionosphere models
Solar protons	Proton flux	Geosynchronous orbit or L1	1 hour	Proton detector on GOES spacecraft, ACE or other upstream monitor	Mean absolute error in log	Ionosphere and radiation belt models
Solar wind	n, P, v_X, B_X, B_Y, B_Z, ram pressure, E_Y, E_Z, Akasofu ε	L1 solar-wind monitor	5 minutes	ACE spacecraft	Mean absolute error	Magnetosphere model
Disturbance departure times from the Sun	Time when disturbance leaves Sun	Earth's surface	N/A	SOHO, Mauna Lea Solar Obs, and other ground locations	Mean absolute error	Thermosphere, ionosphere, and magnetosphere models
Solar wind transit times	Transit time from Sun to Earth	L1 solar-wind monitor	N/A	ACE spacecraft	Mean absolute error	Magnetosphere model

Define a sudden, major change in the solar wind speed to be 50 km/sec within three consecutive five-minute averages. Every five minutes constitutes a separate trial for an algorithm, which is required to predict (at its option) *either* the time (t′) of the next major change, *or* the end time of an interval during which there will be no such changes (t″). If the algorithm yields a t′, a t″ is also assigned, for the purposes of the metric, one half hour earlier than t′. By definition, a t″ is assigned for each trial, whereas a t′ is only assigned if the algorithm predicts an event.

With respect to t′, the algorithm succeeds if the next sudden, major change occurs within plus or minus one half hour of t′. With respect to t″, the algorithm succeeds if no sudden, major change occurs before t″.

The metric is a set of four histograms showing the predictions for t′ and t″ separately for successes and failures. The overall predictive power of the algorithm can be judged qualitatively by examining the ratio of successes to total trials as a function of advance prediction time. Specific quality factors t_{90} and t_{50} are defined as the longest advance prediction time within which the algorithm succeeds in 90% and 50% of the cases.

Figure 2-4. NSWP Solar and Interplanetary Space Weather Metric 1

The highest priority prediction in the Solar-Interplanetary regime is the timing of sudden, major changes in the basic properties of the solar wind, such as wind speed. The useful time-scale of advance prediction starts at about one hour, since a spacecraft at L1 can give one hour of warning to the Earth, and one wants to improve on this.

2.5 Marking Progress with Metrics

Setting up a system of scientific metrics and arranging for their routine evaluation for different algorithms will take both time and effort. However, the effort is clearly worthwhile and represents a necessary step toward establishing major practical benefits from space weather research and clearly marking the most expeditious pathways to operational capability. The metrics will provide an objective measure of progress within the NSWP as well as providing useful input for leadership and management action. The NSWP and its participating agencies should use the metrics to determine where additional funding or new research efforts are most needed to achieve the goals of the Program. A space weather metrics program will also promote scientific advance, by stimulating competition and encouraging algorithm developers to confront observational data in a new way that is both rigorous and objective.

Having reviewed current capabilities and after establishing the metrics framework for assessing progress, it is time to look forward by setting operational goals and determining what needs to be done to reach those goals.

2.6 Operational Goals

The goals of the National Space Weather Program (NSWP) are listed in Table 2-7, which shows the parameters that must be specified and forecast in 15 space weather domains. These were established by the civilian and DOD communities in response to customer operational support requirements. Some of these parameters, the neutral atmospheric temperature, for example, are used to drive forecast models. Others, such as the occurrence of coronal mass ejections, are needed to enhance warning capabilities. This list is subject to change as the NSWP proceeds. It will be reviewed and updated periodically as research improves the physical understanding of space weather and as customer needs change.

2.7 What Needs to be Done

Although considerable progress has been made in the last three years, additional advancements are needed so that forecasting abilities can achieve desired levels of accuracy, timeliness, and reliability. Some limited progress can still be made without improving our understanding of the physical mechanisms that generate space weather. This requires ongoing examination and application of data to develop or improve statistical or empirical models. However, in parallel, as our understanding of the space environment improves, physics-based research models must continue to be developed and modified. These models express what we have learned, serve as a test bed for new concepts, and tell us where we still need more work before our understanding is sufficient to meet operational goals. When our understanding of physical processes is satisfactory, we must place operational sensors in the field and transition our research models into operational models that can run on simple computer systems, assimilate operational data, and produce results within operational timelines.

This overall process is expressed in Figure 2-5 showing the NSWP Roadmap. Research must continue in the three areas of physical understanding, model development, and observations. Crosscutting these areas are the three regions of the space environment—solar/solar wind, magnetosphere, and ionosphere/thermosphere. Chapter 3 summarizes advances made in the last three years and additional research that needs to be accomplished to meet the NSWP goals. Appendix A provides a more detailed description of the research objectives and background information for each region of the space environment. Chapter 4 presents timelines for achieving these research goals.

As research provides sufficiently mature knowledge, the new information is adapted for operational use through a well-planned and well-executed technology transfer (T^2)

Table 2-7. Space Weather Domains and Goals

Space Weather Domain	Goal
Solar coronal mass ejections	Specify and forecast occurrence, magnitude, and duration
Solar activity/flares	Specify and forecast occurrence, magnitude, and duration
Solar and galactic energetic particles	Specify and forecast at satellite orbit
Solar UV/EUV/soft x-rays	Specify and forecast spectral intensity and temporal variations
Solar radio noise	Specify and forecast intensity and variations
Solar wind	Specify and forecast solar wind density, velocity, magnetic field strength, and direction
Magnetospheric particles and fields	Specify and forecast global magnetic field, magnetospheric electrons and ions, and strength and location of field-aligned current systems; specify and forecast high-latitude electric fields and electrojet current systems
Geomagnetic disturbances	Specify and forecast geomagnetic indices and storm onset, intensity, and duration
Radiation belts	Specify and forecast trapped ions and electrons from 1 to 12 R_E
Aurora	Specify and forecast auroral optical and UV background and disturbed emissions, the equatorward edge of the auroral oval, and total auroral energy deposition
Ionospheric properties	Specify and forecast electron density plasma temperature, composition, and drift velocity throughout the ionosphere
Ionospheric electric field	Specify and forecast global electric field and electrojet current systems
Ionospheric disturbances	Specify and forecast sudden and traveling ionospheric disturbances; specify and forecast critical propagation parameters
Ionospheric scintillations	Specify and forecast between 200 and 600 km
Neutral atmosphere (thermosphere and mesosphere)	Specify and forecast density, composition, temperature, and velocity from 80 to 1500 km

process, employing but not limited to the Community Coordinated Modeling Center and Rapid Prototyping Centers. This process is described in Chapter 5. The result will be a series of physics-based models that are coupled to account for the interactions between processes in the three regions of the environment. These operational models will be supported by observations from operational sensors, which again will come from the results of research on sensor requirements and technology, and a planned program of technology transfer.

The suite of new models, supported by deployed sensors, will provide a description of the environment sufficient to support production of appropriately accurate, timely, and tailored products. These products will be designed to address the specific needs of customers—commerce, defense, or society as a whole—at a given location and time. The Air Force will issue tailored products to support DOD. SEC will use the model output to provide warnings and alerts to civilian users and will otherwise provide model output to allow private-sector users to generate their own tailored products.

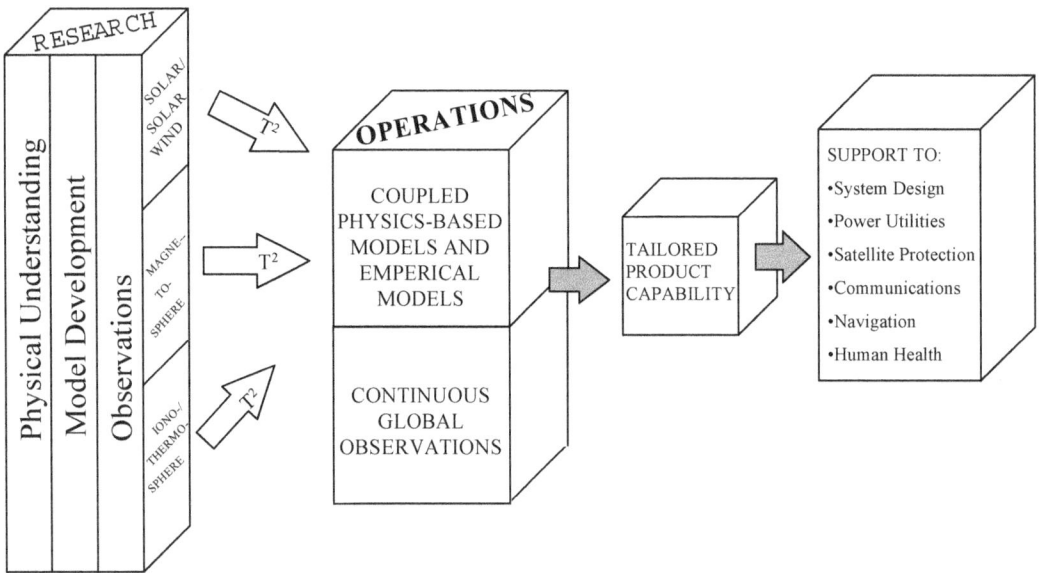

Figure 2-5. NSWP Roadmap

CHAPTER 3

RESEARCH

An operational space weather forecast system requires improved understanding in three broad areas of research: (1) the Sun and solar wind, (2) the magnetosphere, and (3) the ionosphere/thermosphere system. In June 1995, a working group for each of these research areas was assembled to formulate plans for addressing its space weather goals. Each of the three groups was given the goals listed in Table 2-7 and asked to identify what was needed in terms of physical understanding, models, and observations. The fifteen items in Table 2-7 were divided among the three groups, as indicated in Figure 3-1.

Solar/Solar Wind:
- Coronal mass ejections
- Solar activity/flares
- Solar and galactic energetic particles
- Solar UV/EUV/soft x-rays
- Solar radio noise
- Solar wind

Magnetosphere:
- Magnetospheric particles and fields
- Geomagnetic disturbances
- Radiation belts

Ionosphere/Thermosphere:
- Aurora
- Ionospheric properties
- Ionospheric electric fields
- Ionospheric disturbances
- Ionospheric scintillations
- Neutral atmosphere

Figure 3-1. Domains for Space Weather Research

A detailed description of the plan developed by these working groups is contained in Appendix A. Here we provide a summary of the overall plan, including a description of the physical understanding, model development, and observations required in each area.

3.1 Physical Understanding

3.1.1 Background

As outlined in Chapter 2, the National Space Weather Program (NSWP) goals involve the ability to predict the state of the space environment. Hence, basic scientific research must be conducted to improve our fundamental understanding of the physical processes involved.

Beginning with the Sun, a critically important basic research objective is to understand the processes by which coronal mass ejections (CMEs) occur, including the factors that influence their sizes, shapes, masses, speeds, and magnetic field configurations. Equally important is an understanding of solar activity in general. This requires studying how the solar dynamo works and the identification of precursors to solar activity, such as short-term development of active regions and long-term buildup of polar magnetic fields. This involves studying the dynamics of magnetic energy in the solar corona and the role of magnetic fields in the occurrence of flares. It is also important to understand the origins of high-energy solar particles and how they propagate through the interplanetary medium. Similar processes play a role in modulating the fluxes of cosmic rays originating in galactic space. Solar radiation at ultraviolet (UV), extreme ultraviolet (EUV), and soft x-ray wavelengths has a direct effect on Earth's atmosphere. Research in this area is aimed at understanding the variability of the Sun at these wavelengths and how this variability influences the state of the ionosphere and thermosphere. The origin of solar radio noise, which affects communication systems, must also be understood. Finally, the solar wind has a direct influence on the state of Earth's magnetosphere, so it is vital that we understand the processes by which the solar wind is heated and accelerated in the solar corona, as well as the transient perturbations and shocks created by flares and CMEs.

Many space weather applications require knowledge of the particle populations and electromagnetic fields throughout the magnetosphere. This dynamic environment can only be understood by studying the coupling processes between the solar wind and magnetosphere, the transport and energization of plasma in the magnetosphere, the processes involved in magnetic storms and substorms, and the coupling processes between the magnetosphere and the ionosphere. The strong magnetic coupling between the magnetosphere and Earth results in geomagnetic disturbances. The ability to predict geomagnetic disturbances depends on our understanding of the role played by the magnetosphere, ionosphere, and neutral atmosphere (the thermosphere and mesosphere) in modulating the strength of electric currents in space. It is also important to quantify the electrical currents induced in the ground by dynamic currents in the magnetosphere. The magnetospheric radiation belts represent a serious hazard to space systems. Because parts of the radiation belts have been observed to vary significantly, research must be conducted to understand the transport, production, and loss processes that determine the particle flux levels in both quiet and storm times.

Ionospheric properties, including electron density, electron and ion temperature, and composition, are determined by solar radiation, auroral particle impact, and Joule heating. Advances in predictive capabilities in this area depend on our understanding of the formation mechanisms responsible for large-scale and medium-scale electron density structures, and the production, transport, and loss mechanisms associated with these electron density structures. These mechanisms are dynamic in nature and respond to both geomagnetic storms and substorms. The day-to-day variability of large-scale ionospheric features and small-scale plasma density irregularities must be understood to determine their effects on radio wave propagation during quiet and disturbed times. It is also necessary to study the relation between ionospheric irregularities and radio wave scintillation, in particular, the interactions that control the formation and evolution of 10-kilometer to 50-meter electron density irregularities that produce scintillations. Understanding of auroral energy input requires knowledge of the processes that guide, accelerate, and otherwise control particle precipitation, both in quiet times and in times of magnetic storms or substorms. Ionospheric electric fields that drive currents and produce Joule heating must be accurately specified for accurate prediction of ionospheric properties. In particular, it is important to study the small-scale electric field (E-field) structures and the large-scale electrostatic fields and identify the ways in which they couple to the magnetosphere and respond to changes in the interplanetary magnetic field. Further research must be conducted on the process by which high-latitude E fields penetrate to low latitudes. Ionospheric and thermospheric research is strongly coupled, and advancements in the two areas must proceed in parallel. For the neutral atmosphere, basic research needs to be conducted to understand the chemical, radiative, and dynamical processes that act to modify and redistribute energy and constituents throughout the upper atmosphere.

3.1.2 Advances and Work in Progress

The following sections review research advances and work in progress supported under the focused space weather research competitions in 1996 and 1997. Where possible, results already achieved are distinguished from planned research.

3.1.2.1 *Sun and Solar Wind*

Recent research has focused on coronal mass ejections as one of the primary drivers for large geomagnetic storms. This has inspired much work in trying to understand the origins of CMEs and how they propagate into space.

Because coronal mass ejections can produce dramatic increases in the speed and density of the solar wind, it is important to understand the origin of these events. Scientists at the National Center for Atmospheric Research are studying the relationship between different types of solar prominences and CMEs. Preliminary investigations show that certain types of prominences are more likely to yield CMEs, an observation that holds promise in the use of ground-based observations to forecast CME occurrence.

The ability to predict storms on the Sun depends critically on our understanding of the coronal magnetic field structure. A number of models have been developed that compute the coronal field from vector magnetic field measurements at the photosphere. A project being carried out at the University of Hawaii will evaluate the performance of different codes using identical inputs. Both synthetic and real data will be used as input to the models, and discrepancies among the results will be analyzed and discussed in collaboration with the model developers.

Because coronal mass ejections are associated with coronal holes, there is much interest in monitoring and studying the properties of coronal holes. Scientists from Solar Physics Research Institute are investigating coronal holes using He1083 nm full-disk observations made at the National Solar Observatory/Kitt Peak. The study will also make use of photospheric and chromospheric line-of-sight magnetograms to map coronal holes and associated magnetic fields. Knowledge of the variability of coronal holes and their relation to solar activity provides important clues to understanding the short- and long-term evolution of the solar magnetic fields and their extension into space.

Ground-based observation of filaments also may be used for solar forecasting. Investigators from the New Jersey Institute of Technology are developing a procedure based on H-alpha images of the Sun obtained at the Big Bear Solar Observatory to detect filaments and issue an early warning of their occurrence. By creating an automatic system for such detections, the investigators hope to produce a filament index that will reflect the current state of solar activity.

A similar study is being conducted by scientists at Boston College who are studying the occurrence of disappearing filaments and soft x-ray arcades on the Sun. The investigators will use selected disappearing filaments and arcades binned by magnetic polarity and axial orientation to assess the usefulness of these parameters in prediction schemes. The study will show which characteristics of solar mass ejecta are most geoeffective and lead to moderate to large magnetic storms.

Another method for early warning of solar events is being studied by investigators at Hughes STX. Interplanetary type II radio bursts are observed remotely by both the Ulysses and WIND spacecraft. These emissions are generated when CME driven shocks propagate through the interplanetary medium. The investigators have already had some success in using the timing of these radio emissions to estimate the arrival time of the shock at Earth. If validated, this technique will provide a powerful means of estimating shock arrivals many hours in advance of the impact.

Coronal mass ejections produce interplanetary shocks that modulate and accelerate energetic particles. Investigators at Bartol Research are developing theoretical models for this acceleration and the subsequent propagation of the particles to Earth. The models will provide the particle spectrum, fluxes and upper limits for the most energetic particles accelerated at interplanetary shocks.

The ability to predict the effects of solar wind disturbances depends on modeling the solar wind. Satellites at L1 can help, but there is some uncertainty in the propagation of these disturbances from the point at which they are first observed. There have been several studies addressing this issue.

Scientists at SAIC-San Diego are developing a comprehensive, three-dimensional, magnetohydrodynamic model to use remote observations of the Sun to predict the state of the solar wind at Earth's orbit. The code has already been used to determine the coronal magnetic field and heliospheric current sheet structure during the period from February 1997 to March 1998. The model has also been used to simulate the triggering of a coronal mass ejection, including its appearance as seen by a space-based coronagraph.

A model being developed at Johns Hopkins University will use solar wind data obtained by satellites far upstream of Earth's location to calculate the property of the solar wind at Earth's orbit. The results will be analyzed to determine the scale size of solar wind features, the propagation speeds of different structures to Earth, the effectiveness of upstream measurements of the solar wind magnetic field for space weather predictions, and the interaction of the solar wind with Earth's bow shock.

A statistical approach to this same problem is being undertaken by investigators at MIT, who will determine correlation coefficients between plasma and magnetic field properties at the L1 point and at Earth. Because a solar wind monitor at the L1 point is critical to all space weather prediction systems, it is important to understand the relationship between the observed properties of the solar wind at L1 and the variations that occur at Earth's location.

3.1.2.2 *Magnetosphere*

For many years, the Rice University Convection Model has been the mainstay for specifying the magnetospheric response to solar wind inputs. Using the properties of the solar wind impinging on the magnetosphere, the model computes energetic particle fluxes and the convection electric field. In response to the NSWP project solicitation, Rice University scientists are updating the model by incorporating a new open magnetic field model, a magnetofriction equilibrium relaxation technique to achieve particle and field self-consistency, and an algorithm to account for internal plasma sources and losses.

Whereas the Rice Convection Model specifies magnetospheric properties by following the motion of particles in magnetic and electric fields, magnetohydrodynamic (MHD) models solve self-consistently for the particle distributions and fields. Scientists at George Mason University are testing the Lyon-Fedder-Mobarry MHD code using a set of simulated solar input conditions. Such validations are essential to assess the practical limitations of MHD models for space weather applications.

An MHD model developed under the High Performance Computing and Communication program, and now continuing under the Knowledge and Distributed Intelligence Program, is being adapted and tested for space weather applications by researchers at the University of Michigan. This code has many features that make it especially promising, including the use of an adaptive grid to handle both large and small-scale processes. The investigators will develop the code using realistic inputs for solar wind forcing and test the results using other models and observations.

One of the most difficult aspects of magnetospheric modeling is accounting for the dramatic changes in configuration that arise from magnetic substorms. Investigators from UCLA are using an MHD code to study key aspects of substorms. They plan to use real solar wind data as input and compare the model results with magnetospheric and ionosphere observations.

Another group at UCLA is using a combination of MHD simulations and single particle tracing to examine the processes by which energetic particles from the solar wind enter the magnetosphere during storms.

Another validation and testing program for the Lyon-Fedder-Mobarry MHD model is being conducted at Dartmouth University. A large number of solar wind intervals will be selected as input to the MHD model. They will be chosen to include periods of magnetic storms, substorms, and high solar wind dynamic pressure in order to exercise the behavior of the model under a wide range of circumstances. This group is also planning on running a Beowulf cluster to predict the magnetospheric state from real-time solar wind data.

Because of the many satellites in geosynchronous orbit, it is essential that models accurately predict the energetic particle environment at these locations. Researchers at Los Alamos National Laboratory are collecting data from energetic particle detectors on geosynchronous satellites spanning the years from 1989 to the present. These measurements will be used to develop a model of the particle environment in terms of mean fluxes, maxima, minima, and standard deviations.

Researchers at Los Alamos are also working with scientists at Rice University to incorporate a data assimilation capability to the Rice Convection Model. Eventually this capability could be transitioned to the operational Magnetospheric Specification Model as a real-time capability.

Many magnetospheric properties are tied to magnetic indices that have been measured continuously for several decades. It is important to accurately associate these magnetic indices to physical properties of the magnetosphere so they can be used as proxies for quantities that are difficult to measure directly. A study conducted by UCLA scientists will define a better Dst index, create a quick-time index for scientific investigations, and provide insight into the way Dst responds to solar wind input.

Another project to develop more physically meaningful indices is being undertaken at The Johns Hopkins University Applied Physics Laboratory (JHU/APL). These investigators will use data from DMSP and NOAA satellites from 1984 to the present to statistically rate different parameters such as the location of the inner edge of the plasma sheet at midnight, plasma pressure, and conductivity. The relationship between the elements of the database and the geosynchronous field, convection, and substorms will also be investigated.

Other scientists at JHU/APL are using the SuperDARN radars to determine the dayside magnetic merging rate by measuring the transport of magnetic flux across the dayside merging gap in the ionosphere. Work is also underway to extend this technique to include reconnection on the nightside.

Several models have been developed to study the temporal behavior of magnetospheric indices in response to solar wind inputs. These models provide a means to predict the state of the ionosphere from a given time series of input data. Scientists at the University of Maryland are taking these models a step further by constructing models capable of predicting regional properties. The investigators will make use of ionospheric data, ground magnetometers, and solar wind measurements to develop a multivariate model of the dynamical behavior of the different components of geomagnetic activity.

Several different groups are examining the processes by which electrons in the radiation belts are energized during magnetic storms. A group at Rice University is using new mathematical techniques to examine the problem from a theoretical standpoint. A group at the University of Colorado is using a large database of satellite measurements to determine where the maximum phase-space density is located. That location is likely to be the source of the electrons injected into the outer radiation belt.

3.1.2.3   *Ionosphere/Thermosphere*

There are many different types of ionospheric models. These have arisen as a result of different approaches to the problem, or in response to specific requirements. Some models predict the large-scale ionosphere. Some attempt to account for small-scale structure. Others concentrate on specific geographic areas. The goal of the NSWP is to develop a single overarching model for all applications. However, such a model may be composed of several types of interacting models, including physics-based models, empirical models, and data-driven assimilative models. The following projects reflect a multi-pronged approach to ionospheric model development.

A University of Southern California project aims to develop a data assimilation model based on a physics-based model of the ionosphere and measurements of total electron content determined from GPS satellite signals. Tomographic techniques based on GPS signals received at a single station work well only when the ionosphere is relatively unstructured. When combined with a physics-based model, these tomographic determinations can be made more accurately. The investigators will also study the benefits gained from using data from a satellite-borne GPS receiver in low Earth orbit.

This approach has the potential to provide a means to specify the global state of the ionosphere on a continuous basis.

Another study based on GPS signals is being conducted by University of Colorado scientists. This study makes use of GPS signals received in low Earth orbit by the TOPEX/POSEIDON satellite. The research will concentrate in particular on the ionospheric effects produced by large magnetic storms. A second award to University of Colorado scientists will lead to the development of a mid- and high-latitude ionospheric storm-time correction map. If such a correction map is successful, it can be used in conjunction with empirical ionospheric models to better specify the changes that occur during magnetic storms.

Because the topside ionization can contribute 70 to 80 percent of the total column density, it is important to understand how the electron density profile behaves above the F region peak. This is the topic of a study conducted by scientists at Hughes STX who will use data from topside sounders on satellites, incoherent scatter radars, and in situ measurements of electron density to develop a topside ionosphere model. The model will provide a global specification of the topside ionosphere as a function of UT, local time, season, solar cycle, and magnetic activity.

Although several models have been successful in reproducing the large-scale behavior of the ionosphere, the specification and prediction of ionospheric irregularities still poses a serious problem to modelers. The NSWP has supported several projects whose aim is to better understand these irregularities, particularly at the equator and high latitudes where their effects produce serious problems in communication and navigation systems.

A project being undertaken at SRI International involves the deployment of a meridional chain of ionospheric sounders spanning the magnetic dip equator and referred to as the WestPac Chain. The data will be used to study the development of equatorial spread F, including its spatial structure and dependence on ionosphere properties in the conjugate E regions.

Equatorial spread F is being studied in a similar manner by scientists at the Space Environment Corporation. Using data from the WestPac chain of ionospheric stations, the investigators will examine the parameters that control the onset of the Rayleigh-Taylor instability. The observations will be assimilated into models to identify the driving mechanisms and establish the existence of possible precursor events for equatorial spread F.

Another study of equatorial spread F is being conducted at Utah State University using data from the Jicamarca Radar in Peru, along with ionosonde and satellite measurements. The measurements will be used to determine the height-dependent ambient electrodynamic conditions immediately prior to the occurrence of spread F, the location of the initial unstable layer, and its temporal and spatial evolution for different seasons and flux conditions. The combined data set will also allow the investigators to study the role of meridional neutral winds and magnetic declination on spread F formation. When the initial conditions for spread F have been determined, the results will be merged with a global ionosphere model.

A theoretical study being undertaken at Cornell University also strives to achieve a predictive capability for bottomside and equatorial spread F. This study makes use of numerical simulation modeling of the F layer along with observational data from the Jicamarca radar to formulate a theoretical basis for the observations.

A different approach to the prediction of equatorial spread F is being studied by scientists at the University of Texas at Dallas. This involves the development of a neural network using input data from the DMSP satellite and various ground-based observing systems. Historical data are used to determine the combination of circumstances likely to lead to equatorial spread F development. Once the technique has been validated, the model will be examined in detail and contrasted with the output of models based on fundamental plasma physics in order to identify the most critical driving elements of such models.

High latitude ionospheric structure is being studied by Boston College scientists using the Global Thermosphere Ionosphere Model. The model will be run with varying inputs to account for the spatial distribution of ionospheric plasma enhancements in the polar regions, how they form, and how they move in the convection electric field. This study will help to identify the most important processes that need to be included in physics-based models of the high latitude ionosphere.

A study of ionospheric irregularities will also be conducted at Utah State University using the Time Dependent Ionospheric Model. The goal is to develop a model which, when interfaced to a time-evolving ionospheric representation, will yield instantaneous instability growth rates that can be tracked as a function of time along the convecting flux tube. If successful, the study will show that small-scale instability processes can be handled within the context of large-scale ionospheric models.

An important element in all global ionospheric modeling is the ability to accurately specify high latitude inputs, including the convection electric field and auroral currents and precipitation. A study being conducted at the University of Texas at Dallas will provide a means to evaluate the accuracy of models of the high-latitude electric potential distribution. The validation will be based on the deviations between the model predictions and the in situ measurements of electric field made by the DMSP satellite. The study will lead to a set of reliability factors for the electric field models for different ionospheric conditions.

Scientists at Johns Hopkins University are using data from the SuperDARN radar array to study how the convection electric field changes in response to solar activity. An IMF dependent model of the high-latitude electric field potential has been derived from these observations. They are also developing a capability to provide a real-time estimate of the high-latitude convection based on real-time data from multiple radars.

Other scientists at Johns Hopkins University are studying the possible use of magnetometer data from the Iridium satellites to determine the location of the auroral oval. Though relatively insensitive, the Iridium magnetometers are able to detect perturbations from field-aligned currents in the auroral zone. The combined information from the 60 or 70 Iridium satellites in orbit could provide a means of specifying the instantaneous location of the auroral oval almost continuously on a global basis. The technique has been validated for several selected passes, but further work is needed to assimilate data from the entire fleet of satellites. Unfortunately, the bankruptcy of the Iridium group leaves the future availability of these data in doubt.

A Boston University project aims to study the response of the ionosphere and ionospheric currents to the semi-annual variation in geomagnetic activity stemming from the geometric orientation of the dipole relative to the Sun. Because the solar wind is coupled to the ionosphere through the magnetosphere, this study will elucidate the physical processes that connect these extended regions of space.

3.1.2.4   Geomagnetic Storm Studies

Several of the awards made in response to the NSWP project solicitation involved studies of geomagnetic storms from their origin on the Sun to their impacts on the magnetosphere/ionosphere/thermosphere system. Early in the NSWP planning, scientists identified a large storm in November of 1993 as an ideal candidate for such an end-to-end study. Led by an investigator at the U. S. Air Force Academy, a series of concurrent studies was conducted examining the detailed observations of this storm, applying the data to various models, and testing the predictions against observed effects.

Measurements made by incoherent scatter radars are an important element in studying the ionospheric effects of storms, but these instruments do not run continuously and operation must be scheduled in advance. An award was made to SRI International to develop a protocol in which the possible onset of a magnetic storm was predicted using solar observations from the SOHO satellite. An early warning system was set up that successfully allowed for the initiation of radar operations at Sondrestrom, Millstone Hill, and EISCAT for the entire duration of several storms in 1997 and 1998. This procedure will greatly enhance the number of magnetic storm data sets for which incoherent scatter radar data will be available.

3.2 Research Model Development

3.2.1   Background

Many different types of models are important for achieving the goals of the NSWP. The ultimate goal is to develop an operational model that incorporates basic physical understanding to enable specification and forecasting of the space environment by following the flow of energy from the Sun to Earth. This coupled system of models is to be constructed by merging parallel models for the solar/solar wind, the magnetosphere, and the ionosphere/thermosphere. In addition to this, several other types of models will be necessary. Any forecast model must begin with a detailed specification of the current state of the system, which is provided either by empirical models or by assimilative models that take in available observations and fill in gaps. Also, during the course of development of the full operational model, other approaches including empirical predictive methods will be developed and tested to ensure that the most efficient and accurate method is used in the final system.

The ability to predict coronal mass ejections and their subsequent effects requires models of the initiation process and three-dimensional magnetohydrodynamic (3D MHD) simulations of the resulting disturbances in the solar wind. Models of particle acceleration in the CME-driven interplanetary shocks are also necessary for predicting the intensity and time of arrival of particle events at Earth's orbit. Models of radio emissions from CMEs are important for optimizing the use of radio noise as a remote sensing tool.

In modeling solar flares, it is necessary to know the magnetic field in the corona. The only method for determining this field is to observe it at the photosphere and use numerical modeling to extrapolate it into the corona. Simulating the flare itself requires 3D models of magnetic reconnection in active regions, including consideration of the processes that determine the distribution and magnitude of resistivity. Models relating to the processes by which solar flares accelerate particles and generate UV, EUV, and x-ray bursts are also necessary for accurate prediction of flare effects.

Models for the solar wind include 3D MHD simulations of the coronal acceleration region and the solar wind extension into interplanetary space. A coupled version of these two models can be used as a proxy for specifying solar wind velocity prior to its expansion into interplanetary space. Because it is likely that interplanetary magnetic field (IMF) data will come from a satellite at the Lagrangian (L1) point, a model is also necessary to predict solar wind and IMF conditions at the magnetosphere as extrapolated from the available information.

To predict solar UV, EUV, and x-ray emissions, it is necessary to develop 3D models of the solar atmosphere and improve the solar spectrum calculations covering all significant lines and bands from atoms, ions, and molecules.

Although there are several models in existence that specify and predict the particles and fields in the magnetosphere, the Magnetospheric Specification Model (MSM) and the Magnetospheric Specification and Forecast Model (MSFM) are in operational use. Both of these models depend on accurate specifications of magnetic and electric fields, and ionospheric conductances, including the effects of auroral precipitation. Continuing development of these models will improve and extend the capabilities of the MSM and MSFM. The MSFM represents only one approach to numerical magnetospheric prediction codes. Another approach incorporates global MHD simulations that self-consistently solve for the plasma distributions and electric and magnetic field configuration. Because MHD simulations do not account for thermal drifts where spatial gradients are strong, a merger of an MSFM-like code and a global magnetospheric MHD

3-11

code may represent an important step toward developing a physics-based predictive magnetospheric model. Approaches that utilize adaptive grids show new promise to resolve features over many scale lengths. A desired byproduct of the magnetospheric model is the specification of currents throughout the magnetosphere and ionosphere system. From this information, other codes can be developed that predict geomagnetically induced currents and the magnetic disturbance indexes derived from them. The MSM and MHD models do not specify the energetic particle populations in the radiation belts. Static models for the radiation belts already exist, but there is an urgent need to develop dynamic models to account for the variations in energetic particle fluxes that are observed during storm conditions.

Approaches to modeling the ionosphere include empirical models based on worldwide data sets, assimilative models that incorporate real-time observations, and 3D time-dependent physical models. The Parameterized Real-time Ionospheric Specification Model (PRISM) is an operational system being used at the 55th Space Weather Squadron to ingest real-time ionospheric data from ground- and space-based sensors and produce electron density profiles. To achieve predictive capabilities, it is important to focus future work on dealing with large-scale and medium-scale structures in a self-consistent manner, and to incorporate the effects of storms and substorms. This may require the development of nested-grid and adaptive-grid models. More realistic boundary conditions must be applied with the eventual goal of developing a fully coupled model that encompasses the mesosphere, thermosphere, and ionosphere using computationally fast, empirical-numerical hybrid models. As with magnetospheric models, the E-field configuration is an important element in accurate predictions of ionospheric behavior. The E-field can be specified, analytically, semi-analytically, or empirically, but in all cases is driven by interplanetary parameters and magnetospheric processes. These models must be able to account for the penetration of high-latitude E-fields to low latitudes, and the coupling to neutral atmosphere winds. Ionospheric structures, such as sporadic E, descending layers, equatorial plasma bubbles, auroral blobs, and polar cap patches, must be accounted for in specifying the state of the ionosphere. From the standpoint of satellite-based communication and navigation systems, it is most important to also include the effects of small-scale irregularities associated with these structures that cause ionospheric scintillations. To be operationally useful, the currently available climatological model that specifies scintillation for any radiowave propagation path at any frequency needs to be driven by real-time data from a network of stations. The ultimate goal is to develop a physics-based model incorporating the processes that lead to structuring at all scale sizes.

Neutral atmosphere modeling efforts focus on numerical Thermosphere-Ionosphere-Electrodynamics General Circulation Models (TIEGCMs) that can self-consistently calculate density perturbations and neutral wind systems on a global, 3D, time-dependent basis from physical principles. These models must continue to be upgraded, validated, and tested. Empirical, semi-empirical, and assimilative models of the neutral atmosphere are also important to specify the starting point for physics-based models.

3.2.2   Advances and Work in Progress

Many research models are being developed under the basic research awards funded in response to the NSWP project solicitation. The bulk of the development of these models, however, is being performed through on-going projects funded by NSF, NASA, the DOD, and NOAA.

The following tables catalog current research models, breaking them out into solar and solar wind models, magnetospheric models, and ionospheric models. However, these divisions are not strict and a number of the models have applications in more than one category.

The first column in each table lists an identification number for the particular model for use in the timeline charts presented in Chapter 4. The second column list the model's name and any acronym associated with it while the third column lists the developers and or the points of contact for each model. Electronic mail addresses and World Wide Web Universal Resource Locator (URL) information are provided where available. The fourth column provides a brief description of the type and purpose of the model. The fifth column provides a mostly objective status of the model based on the inputs to the Committee for Space Weather during the development of this plan. The legend for this column is:

- M   -   Mature model
- D   -   In Development
- UD   -   Useable but under development
- MD   -   Mature, but undergoing improvements

A "mature" model is one that is essentially complete and is not undergoing continued refinement. Useable models provide outputs that can be used by the general scientific community. Models that are neither "mature" nor "useable" are still in a development stage and should only be used with the assistance of the primary investigator. Finally, the last column lists the funding agencies for each model from among the federal agencies participating in the NSWP.

Table 3-1. Solar and Solar Wind Research Models

ID	Model Name/Acronym	Contact information	Type and Purpose	Status	Funding Source(s)
S1	SOLAR2000	W. Kent Tobiska kent.tobiska@jpl.nasa.gov	Solar irradiance from x-ray to visible wavelengths.	UD	NASA, NSF, NOAA
S2	Evolving PFSS Coronal Model	Janet Luhmann jgluhman@ssl.berkeley.edu	Coronal magnetic field structure derived from observed photospheric field.	UD	NSF, NASA, DOD
S3	3D MHD Model of the Corona and Solar Wind	Jon Linker and Zoran Mikic linker@iris023.saic.com	3D MHD simulation of the corona and solar wind using observed photospheric magnetic fields as boundary condition.	UD	
S4	Solar Active Region Evolution and Stability	Stephen Keil skeil@sunspot.noao.edu	3-D MHD simulation of solar active region evolution.	D	
S5	Magnetic Breakout of the Sun's Atmosphere (MagBrst)	Spiro Antiochos spiro@zeus.nrl.navy.mil	Ejection of solar flux.	UD	DOD
S6	Wang and Sheeley Expansion Factor Model (WS Model)	Yi-Ming Wang ywang@yucca.nrl.navy.mil	Predicting solar wind speed at Earth from magnetic file observations of the photosphere.	M	
S7	3D Interplanetary Propagation Model (3D IPP)	Victor Pizzo vpizzo@sec.noaa.gov	MHD simulation of global, time-dependent solar wind flow.	UD	NOAA
S8	Shock Time of Arrival/Shock Propagation Model (STOA/ISPM)	Murray Dryer murraydryer@msn.com	Empirical and 2-D MHD interplanetary shock wave.	UD	NOAA

M = Mature   D = In development   UD = Useable but under development   MD = Mature but undergoing improvements

Table 3-1 (continued). Solar and Solar Wind Research Models

ID	Model Name/Acronym	Contact information	Type and Purpose	Status	Funding Source(s)
S9	3D MHD/Kinematic Time-Dependent Shock Propagation-Solar Wind Hybrid Model (HSEM)	Murray Dryer murraydryer@msn.com	Kinematic and 3D MHD code which extrapolates solar magnetic field and wind speed from source surface.	D	NOAA
S10	Global Bimodal Corona and Solar Wind Model (GBMCSW)	Shi Tsan Wu wus@cspar.uah.edu	Quasi-steady state 2D MHD model of helmet-streamers and coronal hole.	UD	NSF and NASA
S11	Streamer and Flux-Rope Interaction Model (SFRI)	Shi Tsan Wu wus@cspar.uah.edu	2D MHD model of helmet streamers and flux ropes.	UD	NSF and NASA
S12	Interplanetary Global Model for Simulating the Evolution of Dynamic and Magnetic Disturbances in the Solar Wind	Marek Vandas vandas@ig.cas.cz	2.5D and 3D MHD simulation of solar wind structures from the Sun to 1 A.U.	D	
S13	3D MHD model for Interplanetary Shock, Stream/Stream and CME propagation through the Solar Wind (Han-Detman 3D Code) also known as the Interplanetary Global Model Vectorized (IGMV)	Tom Detman tdetman@sec.noaa.gov tdet@noaasel.sel.bldrdoc.gov	3D, time-dependent MHD simulation of solar wind beyond 18 solar radii.	MD	NOAA
S14	Bats R Us	Tamas Gambosi tamas@umich.edu	3D MHD Simulation.	UD	NASA, NSF

M = Mature   D = In development   UD = Useable but under development   MD = Mature but undergoing improvements

Table 3-1 (continued). Solar and Solar Wind Research Models

ID	Model Name/Acronym	Contact information	Type and Purpose	Status	Funding Source(s)
S15	Filament and Coronal Chirality Model	Sara Martin sara@helioresearch.org	Statistical Event Predictor based on pattern recognition.	UD	NOAA
S16	Halo Coronal Mass Ejection Model	David Webb webb@phl.af.mil Chris St. Cyr cst@sclc.nascom.nasa.gov	Statistical Event Predictor based on pattern recognition.	UD	NASA, USAF
S17	Coronal Emissions Patterns (Sigmoids) Model	Richard Canfield canfield@helicity.physics. montana.edu	Statistical Event Predictor based on pattern recognition.	UD	NASA
S18	Solar Wind	Syun-Ichi Akasofu sakasofu@dino.gi.alaska.edu	Simulation of solar wind based on solar conditions.		
S19	Solar Wind	Arcadi Usmanov usmanov@snoopy.niif.spb.su	Simulation of solar wind based on solar conditions.		
S20	Solar Wind	Y. Q. Hu	Simulation of solar wind based on solar conditions.		
S21	Magnetic Flux Rope Model	Peter Cargill p.cargill@ic.ac.uk	MHD simulation of solar wind.		

M = Mature   D = In development   UD = Useable but under development   MD = Mature but undergoing improvements

Table 3-2. Magnetospheric Research Models

ID	Model Name/Acronym	Contact information	Type and Purpose	Status	Funding Source(s)
M1	Shue, et al. Model of Magnetopause size and shape	J.-H. Shue and J. K. Chao	Empirical model of magnetopause size and shape.	M	
M2	Petrinec and Russell [1995] Magnetopause size and shape	S. Petrinec	Empirical model of magnetopause size and shape.	M	NASA
M3	Roelof and Sibeck model of Magnetopause size and shape	E. Roelof Ed.Roelof@jhuapl.edu D. Sibeck David.Sibeck@jhuapl.edu	Empirical model of magnetopause size and shape.	M	NASA
M4	Magnetopause location	J. K. Chao T272362@twncu865.ncu.edu. tw	Prediction of location of magnetopause given IMF and solar wind dynamic pressure.	D	
M5	Tsyganenko Magnetic field model [T96_01]	N. Tsyganenko kolya@ndadsb-f.gsfc. nasa.gov	Empirical magnetic field model based on IMF, solar wind dynamic pressure, Dst index, and dipole tilt angle.	MD	NASA, NSF
M6	Ogino/Walker Global MHD and large-scale kinetic model of solar wind particle entry into the magnetosphere	R. Walker rwalker@igpp.ucla.edu	Global MHD simulation with additional kinetic calculation of particle entry at the magnetopause.	MD	NASA
M7	Equilibrium Tail Model	J. Birn	Self-consistent model of magnetic field and isotropic pressure for the tail (beyond 10 Re). Available in 2-D and 3-D versions.	UD	DOE

M = Mature   D = In development   UD = Useable but under development   MD = Mature but undergoing improvements

Table 3-2 (continued). Magnetospheric Research Models

ID	Model Name/Acronym	Contact information	Type and Purpose	Status	Funding Source(s)
M8	Time Dependent MHD code	J. Birn	Time-dependent resistive MHD code.	UD	DOE
M9	3-D Electromagnetic Particle Model (EMPM)	K.-I. Nishikawa kenichi@rouge.phys.lsu.edu	Global electromagnetic particle simulation of magnetosphere.	D	NSF
M10	Rice Field Model (RFM) also known as the Toffoletto-Hill [1993] model (TH93)	F. Toffoletto toffo@alfven.rice.edu T. Hill hill@alfven.rice.edu URL http://rigel.rice.edu/~ding/rfm .html	Theory based model of magnetospheric magnetic and electric fields.	UD	NSF, NASA
M11	Rice Convection Model (RCM)	R. Wolf wolf@alfven.rice.edu	Inner-magnetosphere model.	UD	NSF, NASA
M12	Magnetospheric Specification Model	R. Wolf wolf@alfven.rice.edu	An operational version of RCM.	M	DOD
M13	Fully-adiabatic response model for relativistic electrons	A. Chan aac@landau.rice.edu	Physics-based radiation belt flux mapping model.	UD	NSF, DOD
M14	Substorm electron injection model	A. Chan aac@landau.rice.edu	Hybrid test particle calculation of energetic electrons using the Birn and Hesse MHD model.	UD	NSF, DOE
M15	Hydromagnetic wave-particle interaction model	A. Chan aac@landau.rice.edu	Gyrocenter test particle calculation of wave-particle interactions in inner magnetosphere.	UD	NSF

M = Mature   D = In development   UD = Useable but under development   MD = Mature but undergoing improvements

Table 3-2 (continued). Magnetospheric Research Models

ID	Model Name/Acronym	Contact information	Type and Purpose	Status	Funding Source(s)
M16	Linear Prediction Filter (LPF) model of relativistic electron flux at geostationary orbit	Dan Baker baker@lynx.colorado.edu	Prediction of "killer electrons" at geostationary orbit given solar wind speed at 1 AU.	M	NASA, DOE
M17	UCLA Global Geospace Circulation Model (UCLA-GGCM)	J. Raeder jraeder@pallas.igpp.ucla.edu URL http://www-ggcm.igpp.ucla.edu/gem-ggcm-phase1	Global MHD simulation of magnetosphere using solar wind speed and density, IMF and F10.7 flux data as inputs.	UD	NSF, NASA
M18	Ogino S-M Coupling model	T. Ogino ogino@stnet1.stelab.nagoya-u.ac.jp	Global MHD simulation of magnetosphere.	UD	
M19	Dartmouth-NRL-UMD MHD model	J. Lyon, J. Fedder	Global MHD simulation of magnetosphere using solar wind speed and density, IMF as inputs.	UD	NSF, DOD
M20	BATS-R-US magnetospheric simulation model	T. Gombosi	Global MHD simulation of magnetosphere using solar wind speed, density and IMF as inputs.	UD	NSF, NASA
M21	Integrated Space Weather Prediction Model (ISM)	Bill White bwhite@mrcnh.com	Integrated 2-fluid MHD model of magnetosphere with coupling to physics based ionosphere/thermosphere model.	UD	DOD

M = Mature   D = In development   UD = Useable but under development   MD = Mature but undergoing improvements

Table 3-3. Ionospheric Research Models

ID	Model Name/Acronym	Contact information	Type and Purpose	Status	Funding Source(s)
I1	Polar Cap Potential Drop Model [1981]	P. H. Reiff	Empirical model of potential drop across the polar cap based on solar wind velocity and IMF.	M	
I2	Heppner-Maynard-Rich convection model	F. Rich rich@plh.af.mil	Empirical model of ionospheric convection based on IMF.	M	
I3	Izmiran Electrodynamic Model (IZMEM)	V. Papitashvili papita@pitts.sprl.umich.edu	Empirical model of ionospheric convection based on IMF and solar wind speed and density.	MD	
I4	IZMEM/DMSP	V. Papitashvili papita@pitts.sprl.umich.edu F. Rich rich@plh.af.mil	Empirical model of ionospheric convection based on IMF and solar wind speed and density.	D	DOD, NSF
I5	Weimer Electric Potential Model (W96)	D. Weimer dweimer@mrcnh.com	Empirical model of ionospheric convection based on IMF.	MD	NSF
I6	Space Weather Ionospheric Forecast Technologies (SWIFT)	N. Maynard nmaynard@mrcnh.com	Empirical prediction of ionospheric potential patterns, currents and Joule heating driven by L1 solar wind data.	D	NSF, NOAA
I7	APL ionospheric convection model	J. M. Ruohoniemi Mike.Ruohoniemi@jhuapl.edu	Empirical model of ionospheric convection based on IMF.	UD	NSF
I8	Kamide-Richmond-Matsushita (KRM) model	Y. Kamide kamide@stnet1.stelab.nagoya-u.ac.jp	Derivations of ionospheric convection and currents from magnetometer data and conductivity model.	M	
I9	Assimilative Mapping of Ionospheric Electrodynamics (AMIE)	A. Richmond	Derivation of ionospheric convection and currents based on a conductivity model and inputs from magnetic and electric field measurements.	MD	NOAA, NSF

M = Mature D = In development UD = Useable but under development MD = Mature but undergoing improvements

Table 3-3 (continued). Ionospheric Research Models

ID	Model Name/Acronym	Contact information	Type and Purpose	Status	Funding Source(s)
I10	APL Spherical Harmonic Expansion of Polar Cap Potential	J. M. Ruohoniemi Mike.Ruohoneimi@jhuapl.edu K. Baker Kile.Baker@jhuapl.edu	Global Polar Cap potential and ionospheric conductivity derived from radar electric field measurements.	UD	NSF, NASA
I11	Weimer field-aligned current model	D. Weimer dweimer@mrcnh.com	Empirical model of field-aligned currents.	D	NSF
I12	Millstone Hill Electric Field Model	J. Foster jcf@haystack.mit.edu	Empirical electric field model derived from incoherent scatter radar measurements.	MD	NSF
I13	Fejer-Scherliess storm-time zonal electric field model	B. Fejer bfejer@cc.usus.edu	Empirical low-latitude electric field model.	UD	NSF
I14	Scherliess-Fejer quiet-time equatorial vertical drift model	B. Fejer bfejer@cc.usus.edu	Empirical equatorial model of vertical plasma drifts.	UD	NSF, NASA
I15	International Reference Ionosphere [1995] (IRI 95)	D. Bilitza bilitza@nssdc.gsfc.nasa.gov	Empirical model of ionospheric electron density, electron temperature, ion temperature, ion composition.	M	
I16	Mass spectrometer Incoherent Scatter Radar Model of Thermosphere (MSIS)		Empirical model of thermospheric temperature, composition and mass density as function of time, F10.7 flux and magnetic activity.	M	
I17	TIME-GCM Thermosphere-ionosphere-mesosphere electrodynamic general circulation model	R. Roble	Physics-based simulation of the thermosphere, ionosphere and mesosphere.	UD	NSF, DOC

M = Mature   D = In development   UD = Useable but under development   MD = Mature but undergoing improvements

Table 3-3 (continued). Ionospheric Research Models

ID	Model Name/Acronym	Contact information	Type and Purpose	Status	Funding Source(s)
I18	Three-Dimensional Ionospheric Model (TDIM)	Bob Schunk and J. Sojka	Physics-based simulation of the ionosphere using MHD magnetosphere model as input.	UD	NSF
I19	Thermosphere-Ionosphere Nested Grid Model (TING)	T. Killeen	Physics-based but semi-empirical model of coupled thermosphere ionosphere at high latitudes.	UD	
I20	Coupled Thermosphere Ionosphere (CTIM), Coupled Thermosphere, Ionosphere, Plasmasphere (CTIP) and Coupled Thermosphere, Ionosphere, Plasmasphere with self-consistent Electrodynamics (CTIPE)	T. Fuller-Rowell tjfr@sec.noaa.gov	A hierarchy of global, physics based models of the thermosphere, ionosphere and plasmasphere.	MD	NSF, NASA, DOD
I21	Sheffield University Plasmasphere-Ionosphere Model (SUPIM)	G. Bailey	Physics-based.	M	
I22	Field Line Interhemispheric Plasma Model (FLIP)	P. Richards	Physics-based, 1-D, time-dependent model of ionospheric and plasma sphere.	M	
I23	Ionospheric Irregularity Model	J. Sojka	Physics-based model of ionospheric plasma density irregularities. Driven by the TDIM ionospheric model.	D	NSF, DOD

M = Mature   D = In development   UD = Useable but under development   MD = Mature but undergoing improvements

Table 3-3 (continued). Ionospheric Research Models

ID	Model Name/Acronym	Contact information	Type and Purpose	Status	Funding Source(s)
I24	Coupled Ionospheric Scintillation Model (CISM)	S. Basu	Physics-based model of equatorial scintillation.	D	DOD
I25	Wideband Scintillation Model (WBMOD)	A. J. Coster and S. Basu	Climatological model of ionospheric scintillation.	UD	DOD
I26	Hardy, et al. model of ionospheric conductivity	D. A. Hardy	Statistical model of auroral particle precipitation and conductivity	M	DOD
I27	Wallis and Budzinski model of height integrated conductivities	D. D. Wallis	Empirical model of height integrated conductivities in the ionosphere.	M	
I28	Spiro, Reiff and Maher model of auroral conductances	R. W. Spiro	Empirical model of precipitating electron energy flux and auroral conductances.	M	
I29	Fuller-Rowell and Evans model of height-integrated Pedersen and Hall conductivity patterns	T. Fuller-Rowell or D. S. Evans devans@sec.noaa.gov	Empirical model of ionospheric conductances derived from TIROS-NOAA particle precipitation data.	M	NOAA
I30	Precipitating Electron Model of ionospheric conductances (PEM)	H. Kroehl	Statistical model of particle precipitation and ionospheric conductances.	U	NOAA
I31	Ahn, et al. model of ionospheric conductances	B.-H. Ahn	Empirical model of ionospheric conductances based on ground magnetic disturbance data.	UD	NOAA

M = Mature   D = In development   UD = Useable but under development   MD = Mature but undergoing improvements

3.3 Research Observations

3.3.1   Background

Observations in support of the NSWP include both operational and research-oriented data. Ground-based operational sensors include magnetometers and ionosondes, and sensors for ground-based solar observations at both radiowave and optical wavelengths. Enhancements in ground-based operational sensors include extending the current networks, adding scintillation monitoring systems, upgrading solar observations, and adding a network of solar coronagraphs and interplanetary scintillation monitors. Ground-based sensors for research purposes include the array of high frequency (HF) and incoherent scatter radars, riometers, and optical instrumentation. Future enhancements include the Relocatable Atmospheric Observatory (RAO) designed to fill gaps in the existing chain of incoherent scatter radars.

Space-based sensors for the NSWP must be deployed in many different orbital configurations. Low-Earth orbiting satellites, such as those of the Defense Meteorological Satellite Program (DMSP) and the future National Polar-orbiting Operational Environmental Satellite System (NPOESS), are measuring properties of the ionosphere and thermosphere, as well as the plasma processes at low altitudes along auroral field lines. Highly elliptic, polar-orbiting spacecraft are needed to study ionosphere-magnetosphere coupling. Ideally, they should also carry optical and x-ray imagers for determining the instantaneous distribution of auroral precipitation. Sensors on geosynchronous satellites must continue to monitor the energetic particle populations in the magnetosphere and the solar x-ray emission. The particle monitors on the Global Positioning System (GPS) satellites can also be used to specify radiation belt flux levels. Extremely critical to the success of the NSWP is a satellite at the L1 point between the Earth and Sun to monitor the solar wind. This requirement is being filled currently by the Advanced Composition Explorer (ACE) satellite. Plans are in development to supplement or replace ACE with GEOSTORM, ideally incorporating solar imaging capability in addition to the plasma monitors. The satellites described here will all provide operational data for space weather forecasters. Also important to meeting NSWP objectives is the existing and planned satellites that are part of the Department of Defense and National Aeronautics and Space Administration space missions.

One concern is that many of the existing research satellites are approaching the end of their funded mission lifetimes, and they will effectively be turned off just as we are starting solar maximum. These satellites could provide invaluable research data for understanding the basic science of the Sun, and they could also provide critical operational observations such as solar wind speed, interplanetary magnetic field orientation, etc. It is especially important to continue research observations until they are replaced by operational sensors.

3.3.2   Space-Based Research Observing Systems

The NSWP Implementation Plan timelines included a set of observations expected to be available for space weather research activities. These are distinguished from observations used for operational purposes described in Section 2.2.1. However, many observations are used for both research and operations. This section describes observing systems whose primary mission is research.

3.3.2.1   ACE

The Advanced Composition Explorer (ACE) was launched in August 1997 and was placed in a halo orbit about the sunward Lagrangian point in December 1997. The prime objective of this mission is to determine and compare the elemental and isotopic composition of several distinct samples of matter: the solar wind and energetic solar particles, the local interstellar medium, and matter from nearby regions in the galaxy. The spacecraft also transmits data concerning the solar wind and solar energetic particles that are made available on the World Wide Web by NOAA within about five minutes of real time, thus serving as a monitor of solar weather which can produce geomagnetic storms and other effects upon the Earth. More highly processed data is also distributed on the World Wide Web with a delay of several days.

3.3.2.2   ARGOS

In December 1998 the Air Force Space Test Program (STP) launched the Advanced Research and Global Observation Satellite (ARGOS). This 6000 pound satellite was launched aboard a Delta II rocket into a 0230/1430 sun-synchronous polar orbit at 850 km. ARGOS contains nine separate experiments developed by the Navy, Air Force and Army to test a number of new space technologies and to make measurements of the space environment. Five of the nine experiments are designed for remote sensing of space weather and include:

- High Resolution Airglow/Aurora Spectroscopy (HIRAAS) experiment;
- Global Imaging Monitor of the Ionosphere (GIMI);
- Unconventional Stellar Aspect (USA);
- Coherent Electromagnetic Radio Tomography (CERTO); and the
- Extreme Ultraviolet Imaging Photometer (EUVIP).

These five experiments consist of eight separate instruments including ultraviolet imagers, and limb-scanning spectrographs, an x-ray detector and radio beacon to measure the composition, density, temperature and dynamics of the thermosphere and ionosphere. The HIRAAS experiment contains three limb scanning spectrographs covering the wavelength range from 50 nm to 340 nm at resolutions varying from 0.05 nm to 1.7 nm. The GIMI experiment consists of two ultraviolet cameras with 10-degree square fields of view and 3 arcmin resolution. The two cameras are mounted to gimbals on two axes and are designed to image the dayside and nightside ionosphere. The USA experiment is a large x-ray detector mounted to a two-axis gimbal and is capable of measuring neutral density profiles from the occultation of x-ray sources. CERTO consists of a two-frequency UHF radio beacon and will provide tomographic images of the ionosphere to ground receivers. The EUVIP experiment contains an ultraviolet imager with a 5-degree field of view and three separate filters to image the ionosphere and magnetosphere.

3.3.2.3   *Arizona Airglow Instrument (GLO)*

The GLO Instrument has been flown on multiple space shuttle missions to observe the Earth's auroral and airglow emissions both in daylight and at night. Star tracking capability has been added to the most recent GLO experiments. This allows absolute absorption measurements on the topside of the neutral atmosphere by tracking stars into the limb. Stars are used as calibration sources. Limb tracking removes pointing variations caused by the limit cycle of the shuttle as data are being recorded.

GLO has a set of imaging spectrographs that simultaneously observe a wavelength range from 115 nm to 900 nm with a resolution of ~0.5nm in the UV and visible regions of the spectrum, and with a resolution of ~1 nm in the near IR. The GLO spectrographs consist of five modules, each with its own CCD detector. One spectrograph is unintensified and has been found to be too insensitive to be useful for airglow observations. The CCD detectors in the other spectrographs are coupled to image intensifier tubes and have two gratings each, so that two spectra are imaged on each detector. The instrument is also equipped with a set of 13 imagers: a support imager that observes in the red to near IR and 12 that are intensified and fitted with band-pass filters. The purpose of the support imager is to provide a pointing reference for the spectrographs, which are co-aligned with the center pixel of this imager. The field of view of each of the co-aligned spectrographs is $0.2° \times 8°$, while the field of view of each of the monochromatic imagers is $15° \times 17°$. The support IR imager has a field of view of about $3° \times 5°$.

3.3.2.4   *ASTRID-2*

ASTRID-2 is an advanced auroral microprobe with the primary objectives of making high-quality in situ measurements of the physical processes behind the aurora, and to demonstrate the usefulness of microspacecraft as advanced research tools. Instruments for measuring such space weather parameters as local electric and magnetic fields, plasma density and density fluctuations, energetic ions and electrons, as well as remote imaging of auroral emissions will be on board. ASTRID-2 should also be able to demonstrate the feasibility of making inexpensive multi-point auroral in situ measurements. It was launched in 1998, piggybacked on a Kosmos-3M launcher from Plesetsk, Russia.

3.3.2.5   *CNOFS*

Communications Navigation Outage Forecasting System (CNOFS) is a planned Air Force satellite in a 700-km low inclination equatorial orbit for specifying and forecasting equatorial scintillations that cause outages in communication and navigation systems. Another objective is to understand the fundamental physics governing space weather at low latitudes. For this, the satellite will carry a wide range of sensors that will include an UV photometer, ion density, ion drift, electric field and neutral wind probes as well as a GPS occultation sensor and a multi-frequency beacon. The CNOFS data stream will be ingested into a Coupled Ionosphere Scintillation Model (CISM), which is expected to provide nowcast and 6-hour forecast of the background equatorial ionosphere and the onset and magnitude of scintillations.

3.3.2.6   COSMIC

The National Space Program Office (NSPO) of Taiwan is planning, in cooperation with the University Corporation for Atmospheric Research (UCAR), to build and fly a constellation of eight microsatellites in low Earth orbit to measure profiles of ionospheric electron density and tropospheric density/temperature/water vapor using on-board GPS receivers during GPS occultations. This program is based on the GPS/MET receiver which flew on the Micro-Lab 1 satellite in 1995. The NSPO satellites, called the Constellation Observing System for Meteorology, Ionosphere and Climate (COSMIC), will include a GPS occultation receiver, a UHF radio beacon (for computer ionospheric tomography), and a nadir viewing ultraviolet photometer. All eight satellites will be launched on a single vehicle and maneuvered into separate orbits spanning a variety of local times, latitudes and longitudes at any given moment. Electron density limb profiles are derived from the refraction of the navigation signal from a GPS satellite in higher orbit. The beacon and photometer are used to determine the horizontal gradients in electron density.

3.3.2.7   FAST

The FAST satellite mission (1996) investigates plasma processes occurring in the low altitude auroral acceleration region where magnetic field-aligned currents couple global magnetospheric current systems to the high latitude ionosphere. In the transition region between the hot tenuous magnetospheric plasma and the cold, dense ionosphere, these currents give rise to parallel electric fields, particle beams, plasma heating, and a host of wave-particle interactions. In concert with the ISTP/GGS spacecraft, WIND, POLAR and GEOTAIL, FAST will provide several measurements which are essential to our understanding of "Sun-Earth Connections," including energy input measurements to the upper atmosphere, which are highly desirable for the upcoming TIMED mission, as well as information relevant to the nation's space weather program.

3.3.2.8   GPS Particle Detectors

At present, six of the GPS satellites have detectors that measure fluxes of electrons above 100 keV and solar protons with coarse energy resolution and no angular resolution. These data are used primarily to monitor the health and status of the satellite, but are also used for scientific research. Replacement for the current fleet of satellites will begin in 2001 with the Block 2F series, all of which will have particle detectors. These data may be available in near real time provided a sufficient number of ground stations is available. When every GPS satellite eventually carries a particle detector, global, instantaneous snap shots of radiation belt particle fluxes will be possible.

3.3.2.9   HESSI

The primary scientific objective of the High Energy Solar Spectroscopic Imager (HESSI) Small Explorer mission is to understand particle acceleration and explosive energy release in solar flares. The hard x-ray/gamma-ray continuum and gamma-ray lines are the most direct signatures of energetic electrons and ions, respectively, at the Sun.

HESSI will provide imaging spectroscopy of the hard x-ray continuum and high-resolution spectroscopy of gamma-ray lines in solar flares. HESSI utilized a single instrument which combines an imaging system consisting of rotating modulation collimators (RMCs) with high-spectral resolution, cryogenically cooled germanium detectors covering from soft x-rays (3 keV) to high energy gamma-rays (20 MeV). HESSI is planned for launch in November 2000.

3.3.2.10   IMAGE

IMAGE is a MIDEX class mission, selected by NASA in 1996, to study the global response of the Earth's magnetosphere to changes in the solar wind. IMAGE will use neutral atom, ultraviolet, and radio imaging techniques to: 1) identify the dominant mechanisms for injecting plasma into the magnetosphere on substorm and magnetic storm time scales; 2) determine the directly driven response of the magnetosphere to solar wind changes; and, 3) discover how and where magnetospheric plasmas are energized, transported, and subsequently lost during substorms and magnetic storms. The spacecraft was launched in March 2000. Real-time data transmission through NOAA and the Air Force has been arranged.

3.3.2.11   International Solar Terrestrial Physics (ISTP)

The full observational assets of ISTP have been operational now for several years. The strategic locations of the ISTP/GGS spacecraft, WIND, POLAR, Geotail, and SOHO allow fundamental measurements to be obtained on the flow of energy, mass and momentum from the Sun, through the heliosphere, into the magnetosphere, with eventual dissipation in the Earth's atmosphere. The extended mission plan is to quantitatively analyze the fundamental global characteristics of the Solar-Terrestrial system before and during solar maximum to complement the seminal achievements by ISTP near solar minimum. The planned scenario for the WIND trajectory through early CY2002 consists of four distinct phases: L1 halo, lunar swing-bys, high inclination petal orbits, Earth return trajectory.

> _WIND_. The goals of WIND are to determine the characteristics of the solar wind upstream of the Earth and to investigate basic plasma processes occurring in the near-Earth solar wind. During the first two years of operation WIND was positioned in a highly elliptical orbit utilizing multiple double-lunar swing-by's to remain mainly sunward of Earth with a maximum apogee of 250Re. This was followed by a halo orbit at the L1 point along the Earth-Sun line. Real-time solar wind data were distributed via NOAA to the World Wide Web preceding ACE launch for the few hours of WIND data reception.

> _POLAR_. The goals of the POLAR mission are to measure the entry of plasma into the polar magnetosphere; to determine the ionospheric plasma outflow; to obtain auroral images and to thereby determine the energy deposited into the ionosphere and upper atmosphere. POLAR has an elliptical 2 x 9 Re polar orbit with a period of approximately 18 hours.

Geotail. The goal of Geotail (led by ISAS) is to measure global energy flow and transformation in the magnetotail to increase our understanding of fundamental magnetospheric processes. The mission, which carries fields and particle instrumentation, has two phases. During the initial phase, the spacecraft spent most of its time in the distant magnetotail (maximum apogee about 200 Earth radii) while during the second phase apogee was reduced to 30 Earth radii.

Solar and Heliospheric Observatory (SOHO). As part of the ISTP program, the goal of SOHO is to study the internal structure of the Sun, its outer atmosphere and the origin of the solar wind. The spacecraft carries instruments devoted to helioseismology, remote sensing of the solar atmosphere and in situ measurement of solar wind disturbances one hour before they strike Earth. SOHO is permanently positioned at the L1 point where it enjoys uninterrupted viewing of the Sun (cooperative with ESA).

3.3.2.12   Interplanetary CME Imager

The Solar Mass Ejection Imager (SMEI) experiment is designed to detect and measure transient plasma features in the heliosphere, including coronal mass ejections (CMEs), shock waves, and structures such as solar wind streamers which co-rotate with the Sun. SMEI will be flown on the CORIOLIS satellite as part of the Air Force's Space Test Program, with a planned launch in December 2001.

SMEI consists of three cameras, each imaging a 60 x 3 degree field of view, for a total image size of 180 x 3 deg. As the satellite orbits the Earth, subsequent images are used to build up a view of the entire heliosphere. SMEI will provide measurements of the propagation of solar plasma clouds and high-speed streams, data to forecast their arrival at the Earth from one to three days in advance, and the baseline parameters for space weather environmental forecasting.

Characterization of size, mass, and frequency will serve as input to models describing solar activity events, flux shedding by the Sun, and generation of the solar wind. SMEI data will provide the physics of CME propagation, interaction of CMEs with solar wind streams (acceleration and deceleration mechanisms), compression of the Interplanetary Magnetic Field, and interplanetary shock formation. SMEI will also provide data on solar wind streams and other co-rotating structures.

The measurements made by SMEI will be highly complementary to NASA's Global Geospace Science program (GGS) of the International Solar-Terrestrial Physics Science Initiative (ISTP), as well as the National Space Weather Program. SMEI measurements, when coordinated with the imaging and in-situ experiments on the YOHKOH, SOHO, TRACE, ACE, and Ulysses missions, will greatly enhance the productivity of the GGS mission.

3.3.2.13   Interplanetary Monitoring Platform 8 (IMP-8)

Interplanetary Monitoring Platform (IMP) 8 is a 1973-launched spacecraft with a suite of detectors which continue to measure _in situ_ magnetic fields, plasmas, and energetic particle populations. It is in a nearly circular 225,000-km (35 R_e), 12-day geocentric orbit. IMP 8 is a source of uniquely situated data for use in correlation with data from other magnetospheric missions (most notably ISTP) and deep-space missions (Voyager, Ulysses) and of a single uniquely long data set for long-term variation studies.

3.3.2.14   Living with a Star (LWS)

NASA's new Living with a Star Program (LWS) seeks to advance understanding of solar variability and its effect on life and society. A more detailed description of the program is presented in Appendix B. LWS consists of an observational portion, based on dedicated missions, and a supporting theory and modeling program. The observational portion of LWS is divided into a Solar Dynamics Network and a Geospace Network. The Solar Network is composed of the Solar Dynamics Observatory and Solar Sentinels. The Geospace Network is made up of Radiation Belt Mappers and Ionospheric Mappers. The networks are designed to complement the currently planned suite of spacecraft developed as part of NASA's Solar Terrestrial Probe line. The LWS networks are planned for launches beginning in 2007. Maximum overlap between LWS missions and Solar Probe missions is desired to take advantage of the simultaneous observations available from the multiple spacecraft.

3.3.2.15   MSX

The MSX satellite was launched on April 24, 1996. Its primary mission is to gather information that will aid in the design of missile defense systems. The satellite has several optical instruments acquiring excellent observations to support basic research in atmospheric airglow and aeronomy. The mission duration is four years. All instruments are operating nominally.

3.3.2.16   ØRSTED

The main purpose of the ØRSTED satellite is to provide a precise global mapping of the Earth's magnetic field. Provisionally, collection of data is planned for a period of 14 months. The measurements shall be used to improve the existing models of the Earth's magnetic field and to determine the changes of the field. The variations both of the strong field from inside the Earth and of the weaker, rapidly varying, field resulting from the interaction between the ion/particle streams from the Sun (the solar wind) and the Earth's magnetosphere are included in the studies. Furthermore, the transfer of energy from the solar wind to the magnetosphere and further down to the lower layers of the atmosphere will be studied. All of these studies will benefit not only from the magnetic field measurements but also from the measurements of the flow of energetic particles around the satellite. ØRSTED was launched in December 1998 together with the ARGOS satellite.

3.3.2.17   SAMPEX

The Solar, Anomalous, and Magnetospheric Particle Explorer (SAMPEX) mission studies solar, heliospheric, and magnetospheric energetic particles observed from a nearly polar, low Earth orbit. SAMPEX was launched in July 1992. Objectives are to: 1) determine the global flux levels of the magnetosphere in response to interplanetary inputs, and assess the influence of magnetospheric precipitating electrons on upper atmospheric chemistry, 2) characterize the magnetospheric radiation levels relevant to space weather issues including satellite performance and anomalies, 3) determine the ionization states of solar energetic particles over a broad energy range, and 4) determine the fluxes of trapped interstellar material (anomalous cosmic rays) and their solar cycle dependence.

3.3.2.18   Solar Probe

Solar Probe, which will make the very first measurements within the atmosphere of a star, will provide unambiguous answers to long-standing fundamental questions about how the corona is heated and how the solar wind is accelerated. The spacecraft, which will provide both imaging and in situ measurements, is targeted to pass within 3 solar radii of the Sun's surface.

3.3.2.19   Solar Terrestrial Probes (STP)

The STP program is a series of missions specifically designed to perform a systematic study of the Sun-Earth system. Its major goals are: 1) providing an understanding of solar variability on time scales which range from a fraction of a second to many centuries, and 2) determining planetary and heliospheric responses to this variability. The line began with TIMED and is expected to continue in the near-term with Solar-B, STEREO, Magnetospheric Multiscale, Global Electrodynamic Connections, and Magnetospheric Constellation. The new NASA initiative "Living with a Star" seeks to speed up the development and deployment of these systems. See Appendix B and Section 3.3.2.14 for more information on the initiative.

> *TIMED*. The Thermosphere, Ionosphere, and Mesosphere Energetics and Dynamics (TIMED) mission will investigate the basic energetics and dynamics of the region where the sensible atmosphere transitions to space. Major mission objectives are: 1) to determine the Mesosphere Lower Thermosphere/Ionosphere (MLTI—60 to 180 km altitude) structure, including variations with local time, latitude, and season, and 2) to understand the MLTI balance between diverse sinks and sources of energy. Instruments are to measure density, temperature, and wind fields at a spatial resolution of 10° x 10°, and an altitude resolution of 5 km. Launch is scheduled for October 2000.

> *Solar-B*. The goal of Solar-B (led by ISAS) is to reveal the mechanisms which give rise to solar variability and study the origins of space weather and global change. The spacecraft, which will be placed in polar Earth orbit, will make coordinated measurements at optical, EUV, and x-ray wavelengths and will provide the first measurements of the full solar vector magnetic field on small scales.

> *Solar Terrestrial Relations Observatory (STEREO)*. The goal of STEREO, which is one of the near-term Solar-Terrestrial Probes, is to understand the origin of coronal mass ejections and their consequences for Earth. The mission will consist of two spacecraft, one leading and the other lagging Earth in its orbit. The spacecraft will each carry instrumentation for solar imaging and for in situ sampling of the solar wind.

Magnetospheric Multiscale. The goal of Magnetospheric Multiscale, which is one of the near-term Solar-Terrestrial Probes, is to characterize the basic plasma processes which control the structure and dynamics of the Earth's magnetosphere, with a special emphasis on meso- and micro-scale processes. The mission will consist of six spacecraft, four identical platforms, which will fly in formation in order to determine the three dimensional structure of plasma boundaries, and two smaller satellites which will provide images of the context in which the in situ measurements are made.

Global Electrodynamic Connections (GEC). The science objective of the planned NASA GEC Mission is to establish the role of the ionosphere in the electrodynamic environment of near-Earth space. Within this theme two major thrusts are identified:

- Resolve the mechanisms responsible for electrical interactions within the ionosphere/atmosphere system and for its interconnection with the magnetosphere
- Determine the important spatial and temporal scales for electromagnetic energy transfer and dissipation processes in the ionosphere/atmosphere system.

A mission definition team that has already been selected by NASA will establish the measurement requirements and spacecraft characteristics.

Magnetospheric Constellation. The goal of this mission is to understand the interactions between the localized and time-dependent drivers of magnetospheric dynamics. These processes can only be understood by monitoring the entire system, both locally and globally. Plans for Magnetospheric Constellation thus envision the placement of up to several tens of autonomous micro-satellites into a variety of orbits, each carrying a minimum set of fields and particles instruments.

3.3.2.20   *Ulysses*

Ulysses is a joint NASA-ESA out-of-the-ecliptic investigation of the heliosphere as a function of latitude. Ulysses is for the first time studying in three dimensions the effects of explosive solar processes and the major changes in solar wind and solar magnetic field associated with an active Sun. The current phase comprises a Ulysses Solar Maximum Mission, when the structure of the heliosphere is expected to be radically different from that explored during the pole to pole pass conducted during 1994-95. Ulysses will be part of an unprecedented armada of spacecraft investigating solar maximum, but it will be the only one observing from out-of-the-ecliptic plane.

3.3.2.21   *Yohkoh/SXT*

Yohkoh is a high energy solar physics mission of ISAS in Japan with collaboration of the US and UK. Yohkoh scientific instrumentation includes a Hard X-ray Telescope (HXT), a Wide Band Spectrometer (WBS), a Bragg Crystal Spectrometer (BCS), and the Soft X-ray Telescope (SXT) provided by NASA. Yohkoh was launched in September 1991 and has successfully observed the decline of solar cycle 22, the solar minimum in 1996, and the subsequent rise in cycle 23 activity. The primary scientific objective of Yohkoh is to relate energetic solar flare phenomena and dynamic coronal structures such as CMEs to the changing topology of the solar magnetic field. The extended Yohkoh mission is expected to continue this observational program into CY 2002—at which time an entire 11-year sunspot cycle will have been recorded. Data from all Yohkoh instruments are archived for public access in the GSFC Solar Data Analysis Center and have been available via NOAA.

3.3.3   Ground-based Research Observing Systems

3.3.3.1   Automatic Geophysical Observatories (AGOs)

During the austral summer of 1996-97 NSF deployed and made operational the sixth, and last, of the Automatic Geophysical Observatories (AGO). These are small (8x8x16 ft), low powered (50W) autonomous observatories which operate unattended on the East Antarctic Plateau for periods of about a year. They provide some real-time state-of-health and meteorological data via satellite, but most of the science data are stored on optical discs for annual retrieval by aircraft. The AGOs were originally built for studies of the polar ionosphere and magnetosphere, but are also being increasingly used for other Earth related studies. When combined with data from several of the manned Antarctic stations, as well as AGOs which have been developed by other nations, a wealth of information is becoming available on the high geomagnetic latitude ionosphere. AGOs are providing information leading to a much better understanding of the Earth's response to solar activity. Because of the arrangement of land masses, these very high latitude distributed observations are better done from Antarctica than the Arctic; much of the geomagnetically similar regions in the Arctic are over water. The AGO program is a collaboration of scientists from the US, Japan, and the United Kingdom.

3.3.3.2   Balloon-borne Vector Magnetograph

A balloon-borne vector magnetograph was launched over the Antarctic in the summer of 1995-96. The Flare Genesis Experiment, funded by NSF, incorporated an 80 cm diameter telescope to make measurements of the Sun's magnetic field with very high resolution. The experiment had limited success due to technical problems. Two additional launches occurred in December 1998 and January 1999. Each of the experiments circled Antarctica at an altitude of about 120,000 feet for about two weeks before being parachuted to the ice for recovery. The balloons, supplied and launched by NASA, have volumes of about 30,000,000 cubic feet and can lift payloads heavier than a ton.

3.3.3.3   Coronal Magnetic Field Measurements

The magnetic field of the solar corona can be measured in principle by exploiting three physical effects: gyroresonance radiation, the Zeeman effect and the Hanle effect. At present only the first of these techniques is in use on an occasional basis. A proposed Frequency-Agile Solar Radiotelescope would be able to map the magnetic field strength of active regions at coronal heights on a regular basis. The proposal is only in the discussion stage now. The Zeeman effect has been used to measure the line-of-sight component of the corona above the limb from the ground, and it was also used in the EUV to measure sunspot magnetic fields with the SMM satellite. There are no current proposals or plans to use either the Zeeman or Hanle effects for space observations of coronal fields. The University of Hawaii has recently proposed the construction of a 40 cm aperture coronagraph to explore the possibility of making coronal magnetic field measurements above the limb from the ground using infrared emission lines.

3.3.3.4   Incoherent Scatter Radars

The four U.S.-supported incoherent scatter radars in Greenland, Massachusetts, Puerto Rico, and Peru are all operating for approximately 1000 hours per year. There is a move toward coordinating observations in response to the impending arrival of large geomagnetic storms. This will improve the availability of incoherent scatter radar data during storm intervals. An exciting aspect in the operation of the radars is the development of the Upper Atmospheric Research Collaboratory by the University of Michigan. In April, 1998, this group successfully demonstrated the ability to simultaneously display on a computer work station data from all four of the radars, as well as the SuperDARN network, EISCAT, other ground- and space-based instruments, and concurrently running model output. This collaboratory holds great promise in conducting space weather campaigns, coordinated workshops involving a distributed community of scientists, and educational and outreach activities on a global basis. This project is continuing under NSF's Knowledge and Distributed Intelligence initiative as the Space Physics and Atmospheric Research Collaboratory (SPARC).

Another improvement to the incoherent scatter radar chain involves the JULIA radar modification to the Jicamarca facility. JULIA uses the large Jicamarca antenna, but because it operates at low power, can be run continuously to monitor the occurrence of irregularities in the equatorial ionosphere.

3.3.3.5   Interplanetary Scintillation Monitoring

The interplanetary scintillations (IPS) aspect of remote sensing of propagating solar-generated disturbances has been used several times in real-time situations. Scintillations of distant radio sources are detected by ground-based radio telescopes as a result of solar wind density fluctuations along lines-of-sight that pass the Sun at many heliospheric latitudes and at their closest approach. The experiences noted above (published in Solar Physics by Manoharan et al., 1995, and Janardhan et al., 1996) were conducted jointly by the National Centre for Radio Astronomy in Ootacomund, India (Ooty) and NOAA's SEC staff. This collaboration is continuing with observations made on a daily basis with short periods of maintenance.

Several universities have also been working in this area, including the Nagoya University in Japan, the Lebedev Institute in Russia, Beijing University, Santa Maria University in Brazil, and the Universitas Autonomous of Mexico.

Telemetry signals, passing through the inner heliosphere, from interplanetary spacecraft can also be used (given Deep Space Network) to monitor traveling interplanetary disturbances.

3.3.3.6   Relocatable Atmospheric Observatory (RAO)

This proposed transportable system would provide new capabilities for studying the properties of the Earth's upper atmosphere and ionosphere. This collection of instruments, centered on a state-of-the-art phased array incoherent scatter radar with electronic steerability, would give this observatory unique capabilities in addressing problems in solar wind-magnetosphere-ionosphere coupling and its effects on the global atmosphere. For example, locating the observatory near the north magnetic pole would allow observations critical to our understanding of the way Earth's atmosphere is magnetically and electrically coupled to the solar wind. Other possibilities include sites such as Hawaii and New Mexico near large existing lidar facilities, and Poker Flat, Alaska, near the NASA rocket launching facility. The new observations would complement others made by state-of-the-art facilities around the world as well as those made by an international array of satellite-borne instrumentation.

3.3.3.7   Riometers

Riometers measure the absorption of cosmic radio noise originating from ionization below about 120 km altitude. Because this ionization is produced by auroral precipitation, radiation belt particles, and solar flare protons, riometers are an important diagnostic tool for space weather effects. In the past decade, imaging riometers using multiple-beam, phased-array antennas have been deployed in the polar regions, providing enhanced capabilities in the study of dynamical processes. At present, there are 23 imaging riometers operating globally, 14 in the Antarctic and 9 in the Arctic. Real-time data are available from riometers at Gakona, Alaska, and Sondestrom, Greenland. Real-time data from riometers at the South Pole will be available.

3.3.3.8   Scintillation Network

AFRL maintains an extensive network of stations to perform research on the generation, convection, and lifetime of sub-kilometer scale irregularities by monitoring scintillation of radio signals from communications, weather, and GPS satellites. At high latitudes, 250 MHz scintillation receivers are deployed at Sondestrom and Thule, Greenland and at Ny Alesund, Svalbard. Two of these receivers are collocated with incoherent scatter radars whereas all are clustered with other radio and optical instruments. These measurements are focused on the entry of macroscale plasma structures through the cusp, their mesoscale structuring and transit through the polar cap into the auroral oval.
In the equatorial region, 250 MHz and L-band scintillation measurements are performed using a wide array of receivers dispersed in latitude and longitude to provide data on climatology. They are also used in a clustered form with other radio and optical instruments to provide information on irregularity trigger mechanisms, their structure and motion.

3.3.3.9   SuperDARN

The northern hemisphere SuperDARN radar network is currently undergoing an expansion with the construction of radars in Prince George, British Columbia, and Kodiak and King Salmon in Alaska. The radars at Prince George and Kodiak will provide vector plasma drift measurements over Alaska and western Canada. The radar at

King Salmon, Alaska, will provide line-of-sight plasma drift measurements over western Alaska and eastern Siberia.

In the Southern Hemisphere, a new SuperDARN radar became operational in 1999 in Tasmania. A radar at Kerguelen in the south Indian Ocean is under construction and is expected to become operational in 2000. All the SuperDARN radars operate continuously (except for brief maintenance periods).

3.3.3.10 *Vector Magnetographs*

There are currently several instruments that produce vector field maps of solar active regions fairly routinely: The High Altitude Observatory/National Solar Observatory (HAO/NSO) Advanced Solar Polarimeter, the NASA Marshall Vector Magnetograph, the University of Hawaii/Institute for Astronomy Imaging Vector Magnetograph, the National Astronomy Observatory of Japan Solar Flare Telescope vector magnetograph, and the Hairou, China, filter vector magnetograph.

There are no instruments at present that produce magnetographs for the full solar disk. At least three full-disk magnetographs are under construction. The NSO SOLIS project is building a 50 cm vector spectromagnetograph intended to provide regular 1 arc sec full disk vector field measurements in a period of 15 minutes. It will also provide line-of-sight magnetic field component measurements that refer to the solar chromosphere. A proposal has been submitted to the NSF from HAO to build two more of these instruments to be deployed at other locations to achieve nearly continuous solar coverage. Two instruments are under construction in Japan. At the National Astronomical Observatory, a 30 cm aperture infrared Stokes polarimeter will provide 2 arc second pixel images of the full disk in 17 minutes. At the Hiraiso Solar Terrestrial Research Center, a 30 cm aperture Spectroscopic Polarimetry Telescope has been designed to provide improved measurements of active regions. A vector magnetograph is also under construction at the Instituto Astrofisica de Canarias. The US Air Force Improved Solar Observing Optical Network (ISOON) project will allow for future upgrades of the new ISOON instruments to obtain vector magnetograms. A recent MIDEX proposal to NASA called Hale will allow vector magnetograms of the full solar disk to be made from space.

3.4 Summary

Progress in pursuit of NSWP goals has been and will continue to be enabled by synergistic advances in physical understanding, model development, and observations. The next chapter presents the timelines to carry forward this research and development. A critical challenge for the space weather research effort is to provide the linkage between the broad regions (solar/solar wind, magnetosphere, and ionosphere/ thermosphere) of the Sun-Earth system.

CHAPTER 4

TIMELINES AND NEAR-TERM EMPHASIS

4.1 Timelines

The ability of the federal agencies to develop and improve space weather capability to meet the goals and objectives of the NSWP is dependent on the state of the science, the availability of needed observations, communications and computing capacity, and the budgetary environment to support these efforts. The following timelines have been developed with these constraints in mind but still represent an aggressive schedule that should be pursued. This section is broken down into operational sensors and models and research sensors and models. Each examines the evolution of these systems over the period of the NSWP and beyond.

4.1.1 Operational Sensors and Models

4.1.1.1 Operational Sensor Timeline

The emphasis in the sensor timeline is on obtaining measurements of key physical parameters in the spatial domains of space weather from the Sun to the near-Earth environment. The strategy for obtaining the data as quickly and efficiently as possible combines ongoing operational sensors with the use of real-time data from research satellites and sensors. Recent experience has demonstrated that data from research sensors can be made available for space weather operations, at little additional cost, provided planning and coordination for this is included in the mission definition phase of the program. Transmitters compatible with inexpensive ground-based tracking systems have allowed real-time acquisition of data from the ACE satellite and similar transmitters will be employed with the NASA IMAGE mission. Data made available under similar arrangements from future research satellites can also be most useful to operations. Data from the Solar Heliospheric Observer (SOHO) has proven to be extremely valuable in space weather forecasting even while the mission is still in its research phase. Without additional action, the end of SOHO operations will begin a gap in such data until another research mission is flown or an operational mission can be programmed, funded, scheduled, built, and flown—typically a lag of about 10 years. The sensor timeline (Fig. 4-1) shows several such gaps.

Another aspect of the sensor timeline is the evolution from current to desired sensors. The GOES satellite will have a solar x-ray imager beginning in 2002, but the SOHO experiences have shown that the x-ray imager provides better service if accompanied by an EUV solar telescope and a solar coronagraph. A probable platform for these sensors is a GOES weather satellite at geostationary orbit, but sensors on the GOES are defined until at least 2010. SOHO is not expected to remain in reliable operation through that time. The evolution from initial capability to full capability on solar satellite images in the sensor timeline reflects this gap.

Likewise, energetic particle sensors are available on the GOES and these will be upgraded to include lower energy particles. These have been shown to be valuable for satellite anomaly analysis by the Defense Support Program geosynchronous satellites. GOES will begin flying these sensors in 2002, well after the sensors are dropped from the Defense Support Program.

The coverage of energetic particle measurements in the magnetosphere will not be complete until the Compact Environment Anomaly Sensor (CEASE) program or an equivalent is in place later in the decade. That improvement is reflected in the upgrade of the magnetospheric particle sensor program in about 2008.

4.1.1.2   Operational Model Timeline

Current space weather modeling, especially for operational use, is still in its first or second generation. The NSWP promotes evolutionary upgrading by emphasizing funding for developing improved models and providing for their transfer into operations. Improvements will come in an ongoing series of smaller steps that take advantage of new physical knowledge, new observations, and expanded computing capability. The models will be increasingly linked as successive modules in a physical chain extending from the Sun through the terrestrial atmosphere are developed and validated. Another objective with the operational models is to develop ensembles of models describing the same environment. Forecasts can be made from several models of the same phenomena, providing alternative and complementary views of the various aspects of the environment in question. The timeline in Figure 4-2 illustrates the evolution from first generation models through models that have improved capability to meet most requirements but may still fail in some critical situations. The final, fully capable models are able to meet most all of the requirements and projected needs. This includes critical capabilities to adequately describe unusual or intense situations.

Figure 4-2 reflects the transition timeline for models in space weather operations. The models evolve from no capability, or limited capability, to models fully capable of meeting most of the currently foreseen space weather requirements. Most of the models are not envisioned to be fully capable until the second decade of the new century. Part of the basis for this extended period, in addition to the time needed to develop physical

National Space Weather Program Implementation Plan, 2nd Edition, July 2000

http://www.ofcm.gov/

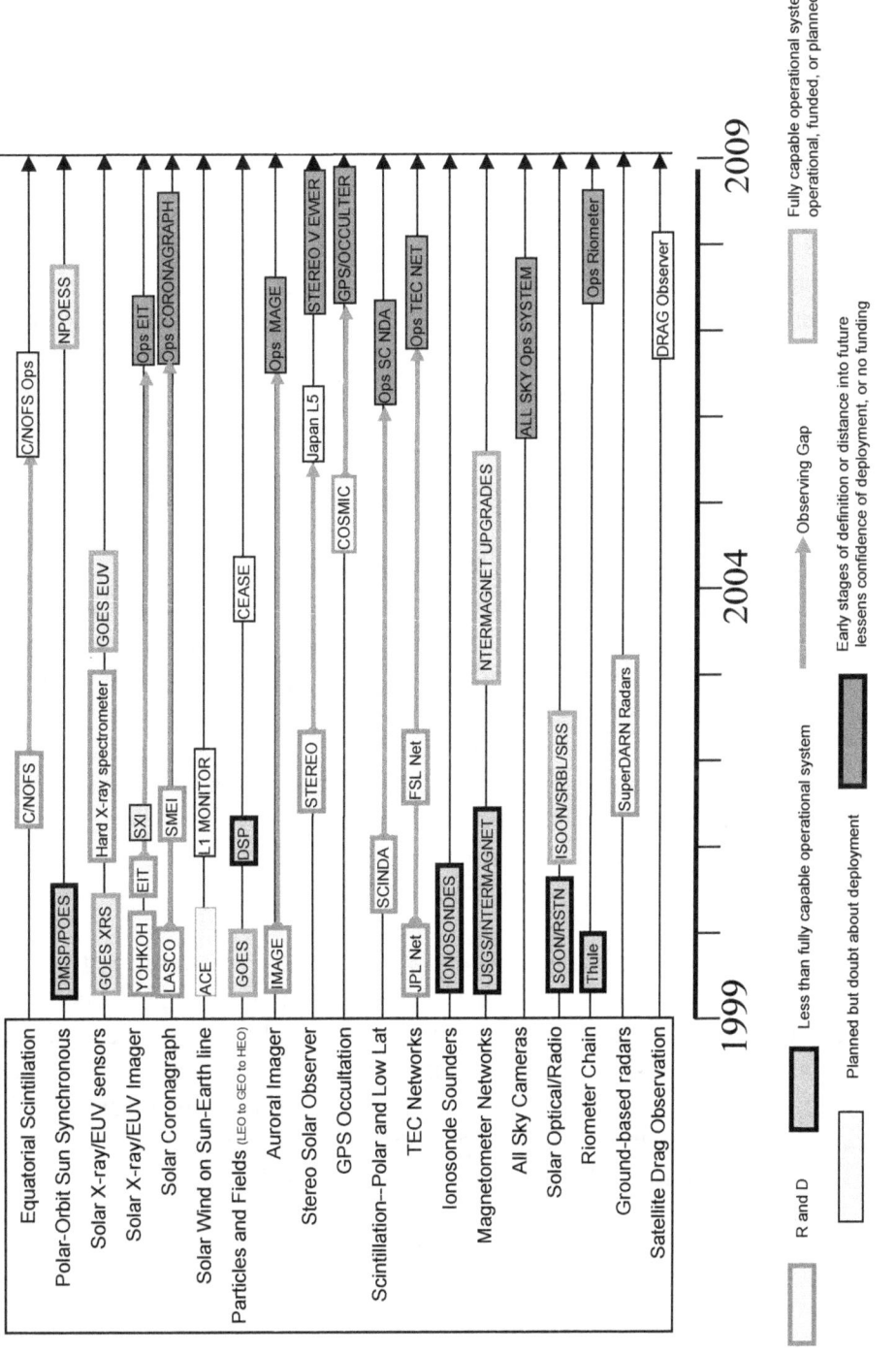

Figure 4-1. Operational Sensor Timeline

4-3

understanding, is an investment strategy that envisions the very large funding need for model development to be spread over two decades. Acceleration of the time to reach full capability can be achieved to some extent by higher levels of funding, but the areas of most benefit need to be carefully balanced against other limitations such as scientific progress and computational capability.

4.1.2   Research Sensors and Models

In many instances it is difficult to clearly distinguish between operational and research sensors and models. This section examines the timeline for sensors not expected to provide routine observations but that are still critical to achieving NSWP milestones and breakthroughs. The research model timeline is much harder to distinguish from the operational timeline because it is expected that successful and usable research models will be transitioned to operations. While the operational timelines were broken out by space weather domain, the research timelines are broken down into three different categories: solar/solar wind; magnetosphere; and ionosphere/thermosphere. The timelines are shown in Figures 4-3, 4-4, and 4-5.

Each figure shows the evolution from 1999 to 2009 of both the research model timeline and the observations timeline, with the models shown on top. The various text boxes are color coded to indicate the current status of each model or sensor system. Green indicates a system that is operating, funded, or has a high confidence of deployment. Yellow indicates models or systems that are planned but confidence in actual deployment is lower. Red indicates models or systems in the early stages of definition and confidence in deployment is low due to their distance into the future, the state of the science, and/or budget issues. Many of the model text boxes contain one or more codes such as I5, M21, or S8. These codes refer to the numbering in the first column of Tables 3-1, 3-2, and 3-3 in Chapter 3 to allow the reader to link specific research modeling efforts to the timelines.

Models in each of the physical domains require information from other domains as input parameters and initial or boundary conditions. Research modeling efforts will increasingly focus on coupling of the various models to ultimately determine the present and future state of the Sun-Earth space weather environment. Underlying the entire process, but not indicated in the figures, is a broad and continuous research effort aimed at acquiring a deep understanding of the physical processes that drive the space weather system. The success of the NSWP depends on a sustained effort in the study of these basic processes.

National Space Weather Program Implementation Plan, 2nd Edition, July 2000

http://www.ofcm.gov/

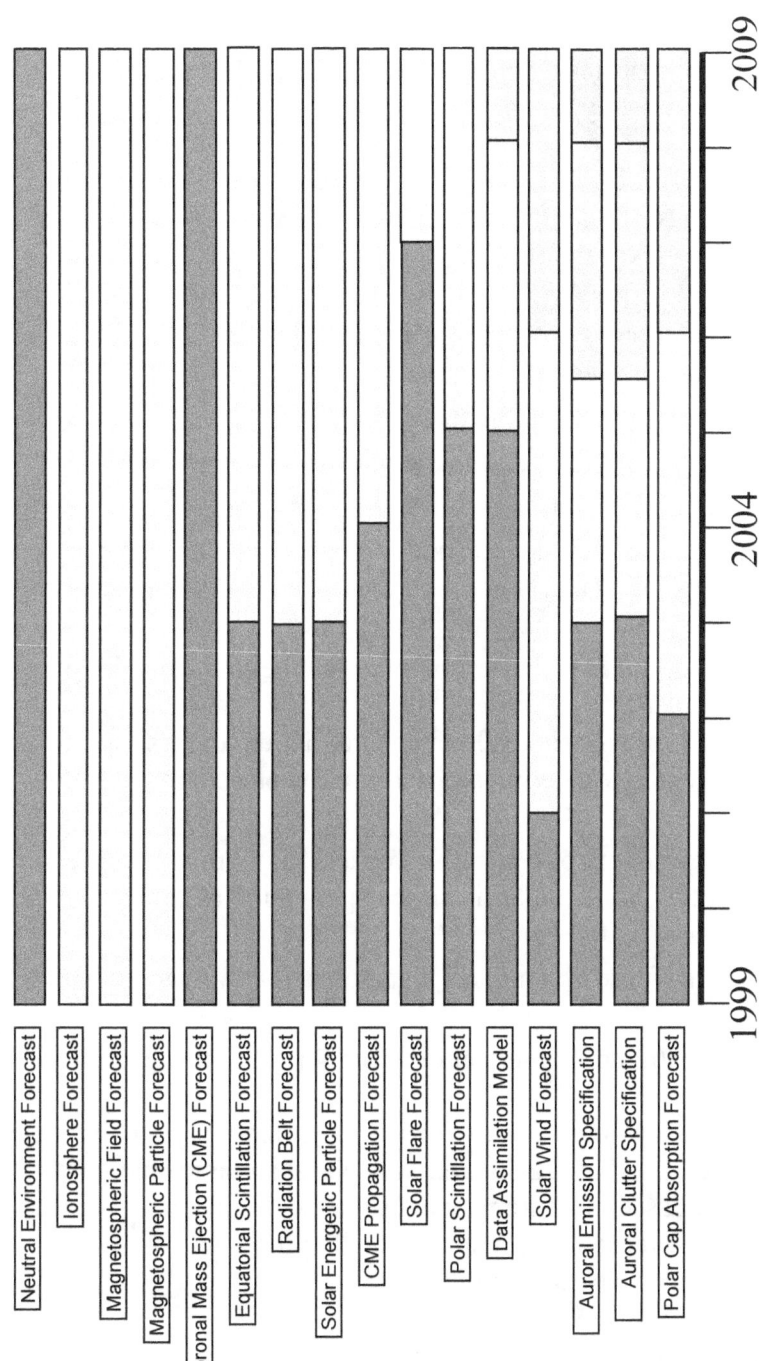

No model or only empirical models whose accuracy does not meet user requirements

Model in use includes some physical understanding but does not meet most user requirements

Evolved capability but model still does not meet some critical requirements

Fully Capable Model

Figure 4-2. Operational Models Timeline

4-5

4.2 Near-term Emphasis

The timelines described above indicate the longer-term milestones that must be achieved to reach the goals of the National Space Weather Program. In addition to these long-term goals, there are also a number of near-term objectives. These near-term objectives will be reviewed and updated approximately biennially.

Physical Understanding. Initially, several broad research areas will be targeted for emphasis. These represent significant gaps in present understanding and need to be addressed early in the Program. They are as follows:

- Understanding and prediction of processes affecting solar activity and solar wind, such as coronal mass ejections and solar flares
- Coupling between the solar wind and the magnetosphere
- The origin and energization of magnetospheric plasma
- The triggering and temporal evolution of substorms and storms
- Improved global ionospheric specification and forecast, including the evolution of ionospheric irregularities, particularly at low latitudes, and with emphasis on those processes affecting communication and navigation systems
- Improved specification of thermospheric dynamics and neutral densities

Models. There is a critical need for model development, validation and testing in all areas of the space weather system. Near-term emphasis will be placed on the following:

- Validation and enhancement of space weather models to improve specification and prediction capabilities, with emphasis on the application of data assimilation techniques
- Continued development of research models that are nearly ready to transition to operations
- Development and application of numerical methods for event forecasting

Observations. In the area of observational requirements, it is important to maintain and continue to improve existing ground-based networks for solar, magnetic, and ionospheric observations. Similarly, space-based measurements must be continued, both for routine observations and to address critical problems in understanding the space weather system. In particular, emphasis should be placed on the following:

- Maintaining ground-based observing systems such as magnetometers and radio and optical solar remote sensing capabilities
- Maintaining sensors on polar-orbiting and geosynchronous satellites
- Progressing with NASA's Sun-Earth Connections satellites and acceleration of their schedules through the "Living with a Star" initiative

- Taking advantage of data from existing satellites such as International Solar-Terrestrial Physics (ISTP), Fast Auroral Snapshot (FAST) and Midcourse Space Experiment (MSX)
- Deploying an operational solar x-ray imaging instrument
- Establishing a new ground-based facility within the magnetic polar cap

Technology Transfer. The challenges of technology transfer for modeling have been addressed very early in the program and the Community Coordinated Modeling Center and Rapid Prototyping Centers (see Chapter 5) show great promise as a means by which research quality models can be prepared for transfer to the operational arena. Progress has been made in technology transfer in the sensor and observational arena through efforts such as the real-time data flowing to the operational centers from the ACE satellite. However, significant challenges remain.

Education. Another important area of early concentration in the Program is customer education. There must be a well-defined procedure by which the scientific community, the forecasting community, and the customers can interact and exchange information. Progress has been made through an NSF grant to the Space Sciences Institute and through ongoing efforts at the SEC and within DOD to educate users and to integrate space weather into normal daily operations. More work needs to be done.

National Space Weather Program Implementation Plan, 2nd Edition, July 2000

http://www.ofcm.gov/

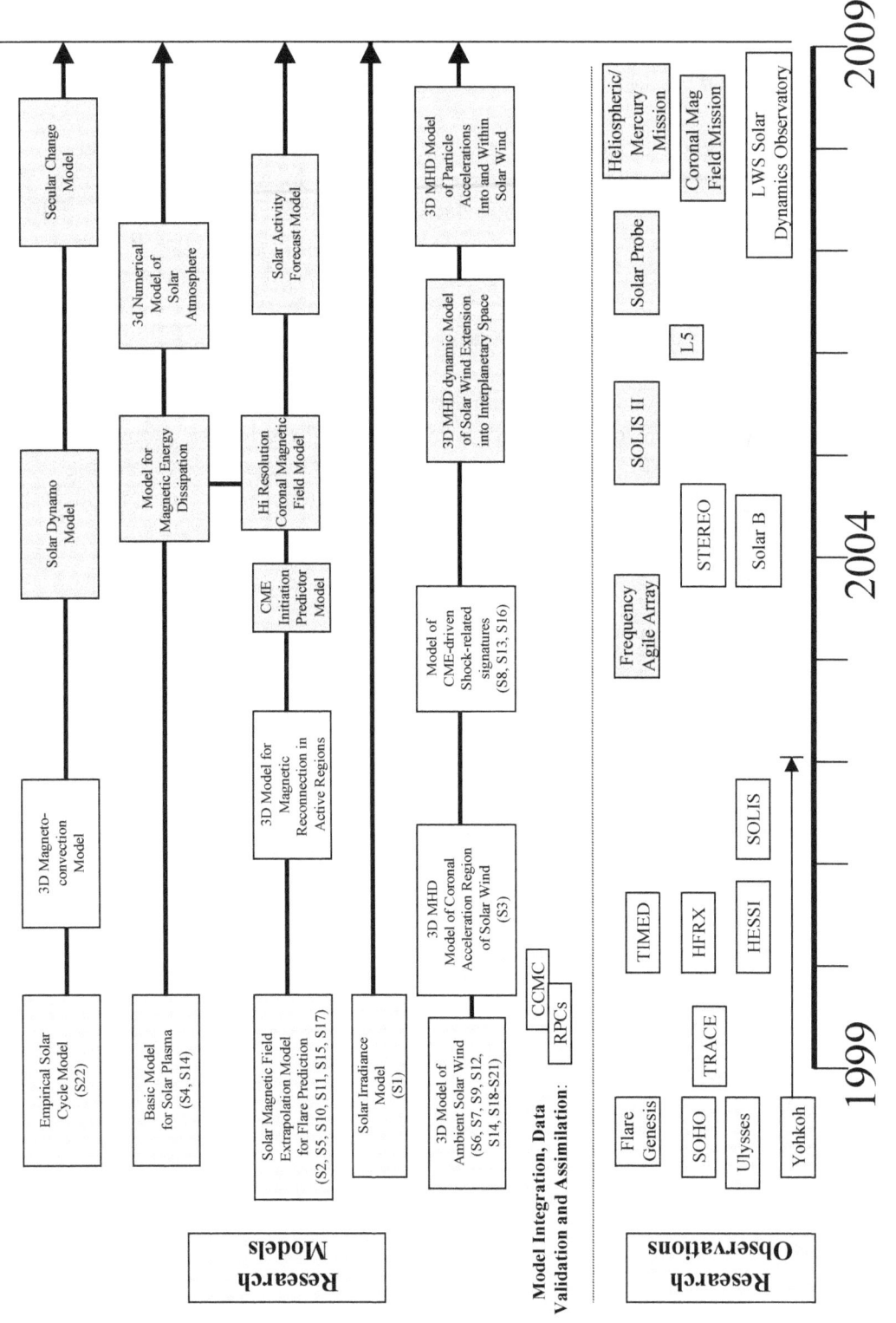

Figure 4-3. Solar/Solar Wind Research Timelines

4-8

National Space Weather Program Implementation Plan, 2nd Edition, July 2000

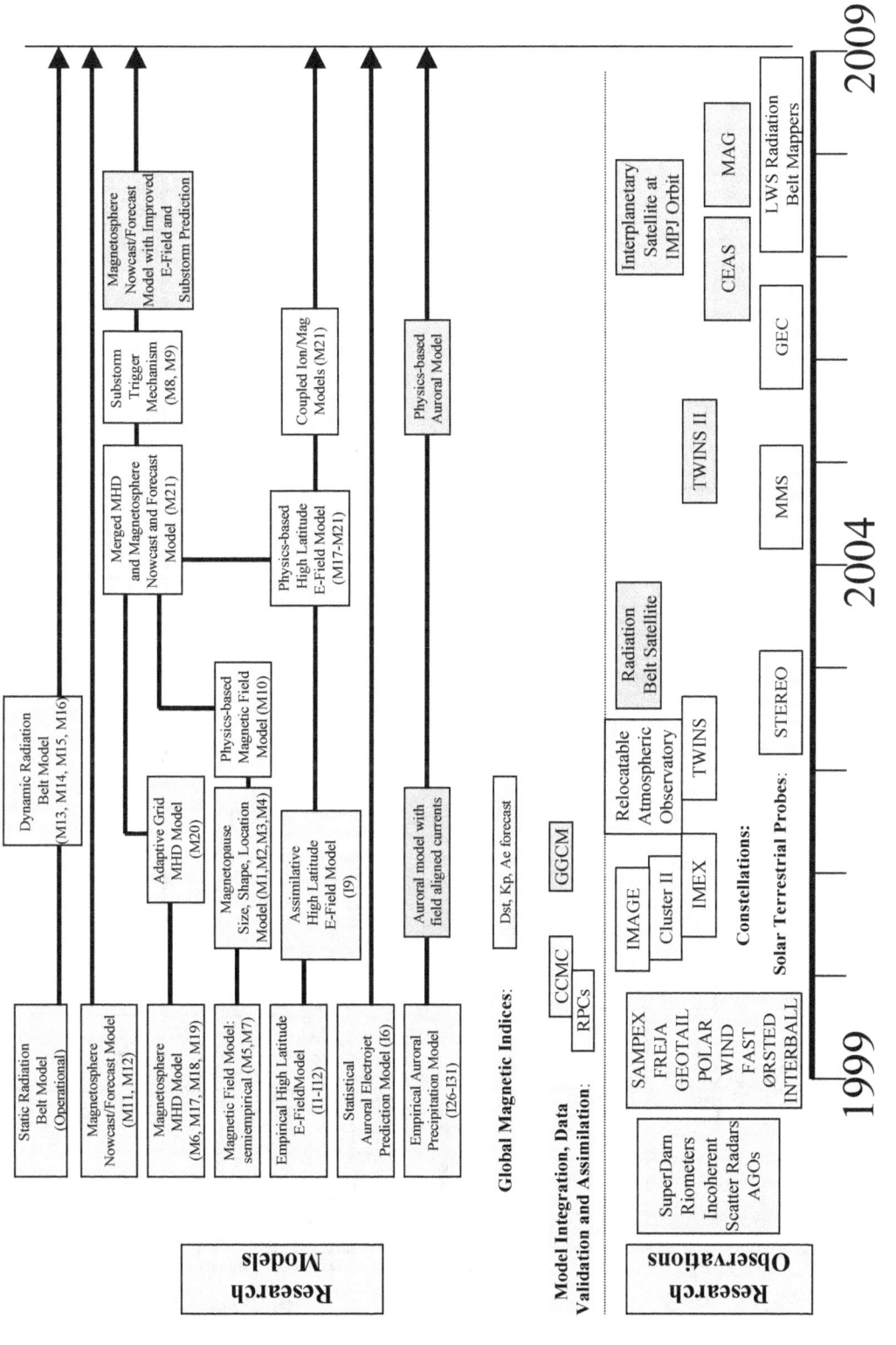

Figure 4-4. Magnetosphere Research Timelines

4-9

National Space Weather Program Implementation Plan, 2nd Edition, July 2000

http://www.ofcm.gov/

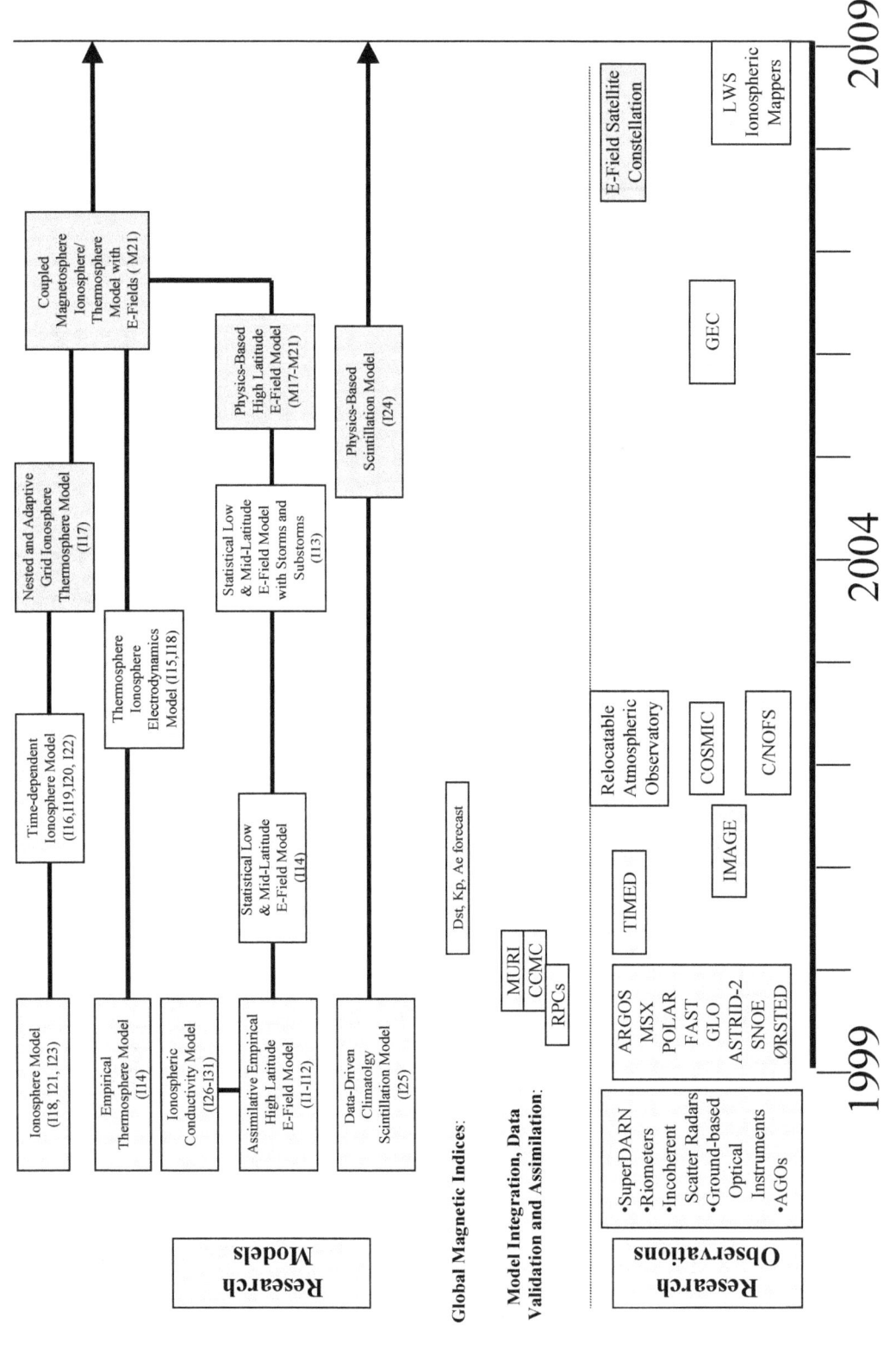

Figure 4-5. Ionosphere/Thermosphere Research Timelines

4-10

CHAPTER 5

TECHNOLOGY TRANSITION AND DATA MANAGEMENT

5.1 Developing Operational Models

Success of the National Space Weather Program (NSWP) requires that the knowledge and models of the space weather system generated by the research community be incorporated into operational models of use to forecasters and/or external customers. The goal of the research model is to demonstrate understanding of the physics appropriate to the bounded system being modeled. Input may or may not include realistic data, and output is typically in the form of scaled parameters that illustrate the behavior of the system under a variety of conditions, mostly idealized. In contrast, an operational model must be able to use existing real data to produce clear-cut results applicable to all types of conditions.

The strict requirements imposed on operational models have made the development process slow and difficult. A high priority in the implementation of the NSWP is the creation of a system to encourage development of appropriate research models and to incorporate research results into operational models quickly and efficiently. At the core of the process are the Community Coordinated Modeling Center (CCMC) and the Rapid Prototyping Centers (RPCs). Figure 5-1 graphically depicts this process. The CCMC provides support and structure to the development of needed research models and transitions them to the RPCs. At the RPCs, immediate feedback is provided to the development team as concepts are tested in a quasi-operational environment. RPCs allow competing methodologies or techniques to be examined quickly, cheaply, and creatively, often generating a product with more capability than originally envisioned.

5.1.1   Community Coordinated Modeling Center (CCMC)

At a Committee for Space Weather (CSW) meeting held at the Office of the Federal Coordinator for Meteorology in March 1998, the Air Force Space Command presented a proposal for a Community Coordinated Modeling Center. This center would provide a place where space science researchers could try out new models currently in the development stage. With further refinement of the concept and unanimous support from

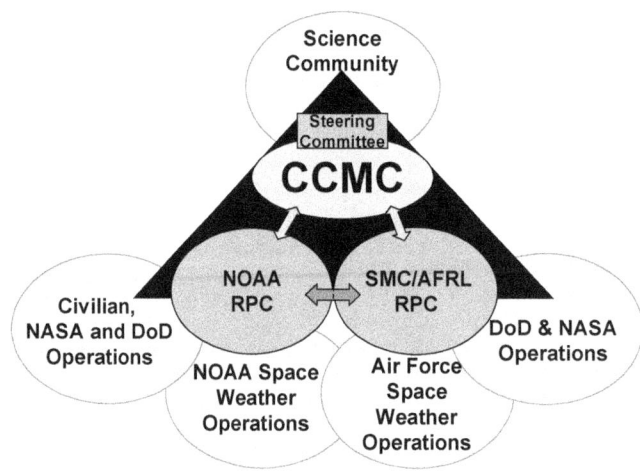

Figure 5-1. Linking the Science Community with Operational Centers

the CSW and space weather community to proceed, a consortium of federal agencies comprised of the Air Force Materiel Command, Air Force Office of Scientific Research, Air Force Directorate of Weather, Air Force Weather Agency, NASA, NOAA, and NSF began development of the CCMC. It will be operational in the year 2000.

The mission of the CCMC is to provide a computing facility to enable, support, and perform the research for next generation space science and space weather models, preparing them for transition to operations through the RPCs. This process is shown graphically in Figure 5-1. However, it should be noted that the CCMC is not necessarily the only conduit for models to move into the RPCs.

The CCMC's goals are to:

- Develop with the community and execute next generation space research models, aiming specifically at space weather needs
- Integrate existing models to cover the necessary range of physical scales
- Provide computational infrastructure for space weather modeling
- Prepare models for transition to rapid prototyping centers
- Make research models developed at CCMC as well as their output available to all in the community
- Support of community research
    - through execution of model runs on request
    - through preparation of visualization and analysis tools
    - through source code dissemination and model output availability
    - through general community-based development
- Perform basic research
    - using existing and newly merged models

- for the development of new models
- in support of space weather goals

In addition to meeting these goals, the CCMC will provide additional benefits such as enhancing scientific understanding, providing broad access to research models, providing effective use of mission data, enabling cross-disciplinary science, and aiding mission conception and design. It is designed to be integrated with the NASA's Sun-Earth Connections Roadmap and the "Living with a Star" program, the National Security Space Architect's Space Weather Architecture Transition Plan, and the NSWP.

The CCMC concept is to meld science and operational requirements and provide a venue for the development and testing of research models with the intent of eventual transition to operations through the Rapid Prototyping Centers. The CCMC operates under the NSWP with an interagency steering committee overseen by the Committee for Space Weather. The Steering Committee consists of individuals from the consortium agencies with assistance from two working groups, one for science and one for operations. The Science Working Group consists of representatives from GEM, CEDAR, SHINE, and ISTP. The Operations Working Group consists of representatives from the Air Force's Space and Missile Systems Center (SMC) and AFRL RPC, the NOAA RPC, and the Headquarters Air Force Directorate of Weather (HQ USAF/XOW).

The Center, located at the Goddard Space Flight Center, Greenbelt, Maryland, supports code adaptation, provides visualization capability and web access, disseminates mature

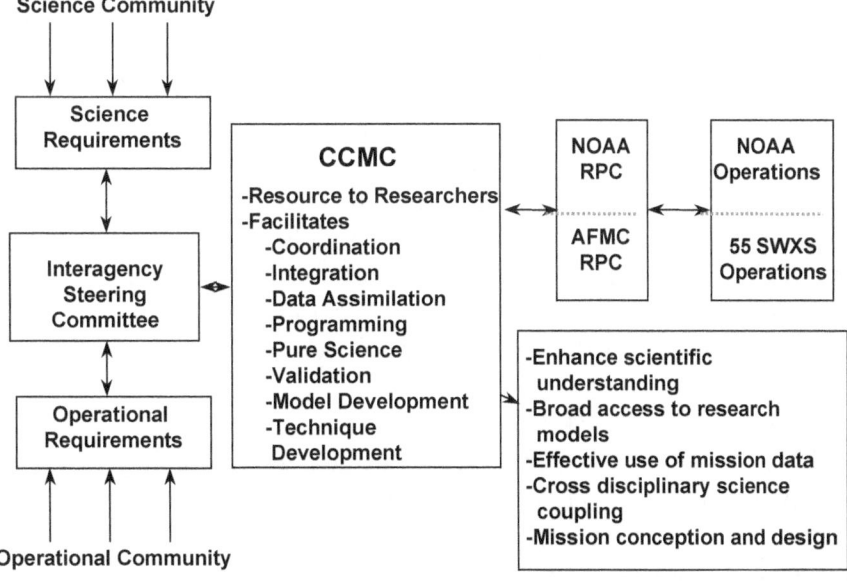

Figure 5-2. CCMC Operational Concept

codes, and performs research in support of CCMC goals. It also provides the front-end computing and access to supercomputer facilities hosted by the Air Force Weather Agency (AFWA), Offutt Air Force Base, Nebraska. Although the core computational capability is massively parallel, this does not preclude the development and validation of smaller models that may be run in a workstation environment. The National Space Science Data Center (NSSDC) provides archived space weather data for use in the CCMC and a real-time data feed into AFWA is planned for the future.

The NSF provides software infrastructure support and, in conjunction with AFOSR, supports research post-doctoral positions in the center. The Air Force and NOAA support the transition to operations process through their respective RPCs and NASA provides the facilities at Goddard Space Flight Center and the supporting technical and research staff.

The CCMC is a novel approach to the development of space weather models and is an example of effective interagency collaboration. The research community has shown considerable interest at GEM, CEDAR, and SHINE workshops and initial model selection has begun.

5.1.2   Rapid Prototyping Centers (RPCs)

The transition of research results into operational capability is incredibly difficult not just for space weather but for virtually every scientific endeavor. However, significant progress has been made in space weather since the initial NSWP Implementation Plan was published in 1997. NOAA has established a Rapid Prototyping Center (RPC) and initiated a Cooperative Research and Development Agreement (CRADA) with Sterling Software to transition the Rice University Magnetospheric Specification Model. The DOD RPC is developing, working with the Air Force Research Lab and Space Battle Lab to create new graphically based, operationally focused space weather products. These products address the DOD operational need for information on ionospheric scintillation impacts to Global Positioning Satellite and Ultra-High Frequency (UHF) satellite communications, High-Frequency (HF) radio communications, and auroral interference with ground radars. When these early successes become operational, the DOD RPC will continue to improve the line of space weather products issued by the 55 SWXS.

5.1.2.1   RPC Concept of Operations

RPCs will publish standards to which models must adhere before they are accepted into the rapid prototyping process. Models must be validated--they must properly represent the physical processes, function correctly within the natural range of input variables, and be coded correctly. Standard structured programming techniques will be required, as will appropriate documentation. Standard computer languages will be specified so that code will be transportable between the hardware platforms and operating system environments under use at the centers. RPCs may also specify output formats to facilitate interaction with in-place visualization tools.

As illustrated in Figure 5-3, the rapid prototyping process uses models matured in the CCMC as well as other research models in conjunction with a realistic data stream. The models are evaluated as necessary to check their ability to produce accurate results under the full range of possible input values. They must be *suitable* to the task at hand, produce *reliable* results, and be easily *usable* to the space weather forecaster. The operational models must deal with momentary or extended data dropouts, different data sampling rates, and rapid rates of change in sensor output that may or may not be noise. Sometimes models must knit together sets of data from very different and non-collocated sensors, extrapolating data, if necessary, or filling in with proxy data. Output errors must be controlled as the model moves forward in time or space, and all model outputs must be self-consistent. The successful operational model should also validate itself against actual measurements and reinitialize itself, or notify operators that errors have become unacceptable. Finally the model should accommodate forecasters with various skill levels, adapt to a variable production demand, and be relatively easy to modify if new data, products, or hardware or software upgrades become compelling.

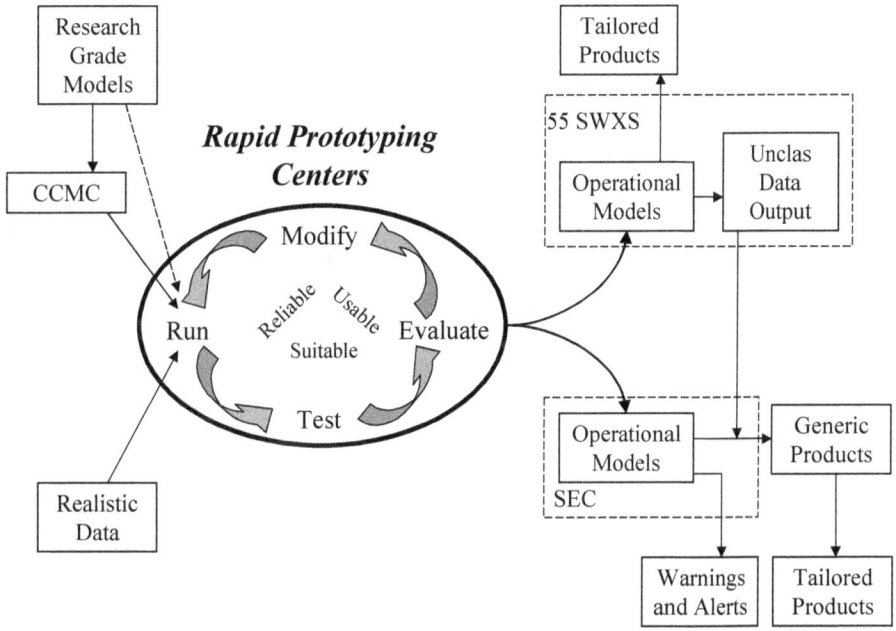

Figure 5-3. RPC Development Process for Operational Models

5.1.2.2   *NOAA SEC Rapid Prototyping Center and Cooperative Research and Development Agreement*

The Rapid Prototyping Center (RPC) is a facility developed at the Space Environment Center to expedite the validation of numerical models and data and to efficiently bring new near-real-time information into operational use. The RPC is developing a competitive and efficient evaluation and transition process to ensure the rapid utilization of the extensive research, modeling, and monitoring efforts directed at the space environment. Its operational output will provide the basis for improved forecasts and for tailored space weather products. The output will be available to the broad user community, in support of SEC operations, the U.S. Air Force space weather operations, commercial space weather concerns, and the research community.

The development of the RPC has been enhanced through a Cooperative Research and Development Agreement (CRADA) with Sterling Software, Inc. SEC entered into this CRADA on June 9, 1997, starting the first public-private partnership to provide space weather services. The SEC/Sterling team has completed an extensive analysis of the requirements for SEC's Rapid Prototyping Center and operational system and has begun the code development and testing of space environment models.

This project has been progressing through a series of iterative stages. The initial stage concentrated on establishing the foundation of computer software that provides the broad functionality of the RPC. The software design follows the object-oriented paradigm and will provide a generic capability for data access, model execution, output verification, output dissemination, and archival.

The second stage involved implementing and providing routine output for a specific space environment model—the Magnetospheric Specification Model. Magnetospheric particle fluxes calculated from the Magnetospheric Specification Model are now available through SEC's Outside User System. The model output is available every three hours as a "routine test product" after being verified by SEC forecasters. The goal of this test product is to allow access to the model output and to obtain feedback, both internally and externally, that will drive further modifications before the model output becomes fully operational. In addition to providing the operational output in support of government services, the output is available to private vendors who will develop value-added products based on the operational output and then market these products to users of space weather services.

The current stage of NOAA RPC development involves expanding the initial software infrastructure to include broader functionality and bringing additional models and data into the transition process and eventually into operations. The RPC data ingest and archival functions will utilize the Information Dissemination System that is under development at SEC to standardize the interfaces that all models will use in development and in operations. Also, the verification and model analysis capabilities will be enhanced over what has been established in the initial stages. As the infrastructure matures, new models are being used to validate the RPC capabilities and to provide additional operational capabilities. Models of solar energetic protons, ionospheric currents, and radiation belt particles are among those currently being considered for near-term transition to operations.

5.1.3 Operationalizing New Models

Research models and analysis techniques that "graduate" from the CCMC and RPC processes will automatically have become incorporated into the appropriate operational space weather centers. The 55 SWXS, which supports military users of space weather data, produces a set of tailored products from the model outputs for specific military needs and provides unclassified data to SEC. These data, along with other data generated at SEC are available for civilian customers. In addition, these data can be used to help industry and, perhaps, universities tailor civilian products in response to specific customer demands. An exciting possibility is the formation of small businesses aimed specifically at tailored-product development such as that envisioned in the new SEC/Federal Data Corporation CRADA signed on May 2, 2000. As larger, more integrated, and more sophisticated models reach operational use, the volume of data feeding these models will grow significantly. Timely development of operational sensors and bridging the gaps between research observational capabilities and operational sensors is imperative.

5.2 Developing Operational Observing Capability

Developing operational models is a high priority of the NSWP. However, for models to mature through increasingly more stringent verification and validation, they will require the necessary observational elements needed to accurately specify and forecast the space environment. These observations may come from dedicated operational observing systems or from research sensors and observing systems adapted to operations. The NSWP also places a high priority on the development of operational observing capability.

Operational sensing systems mature through a similar process, beginning as a research effort to prove the science and technology, then the development and fielding of an operational sensor. Unfortunately, a gap of several years often occurs between the end of a research mission and the subsequent deployment of an operational capability. Every effort must be made to bridge this gap as creatively and effectively as possible. The Advanced Composition Explorer (ACE) spacecraft and the solar wind information it provides is an outstanding example of interagency cooperation and collaboration to employ a research system in an operational mode.

The generally high cost of developing and deploying operational observing systems, especially space-based systems, and continuing budgetary pressures represent a significant obstacle in achieving the NSWP goals. The NSWP will continue to support expanding interagency efforts to bridge the gap between research and operational observing systems and to develop and deploy the operational systems required. As observing capabilities expand, the volume of data produced by these systems will undoubtedly increase, as will the size and complexity of model output based on these observations. An effective data management system must be developed to collect and disseminate space weather data and information.

5.3 Data Management

The NSWP guides improvements in the capability to process, assimilate, analyze, distribute, and archive increasingly complex data sets. Rapid advances in computer and communications technology have opened a realm of possibilities for the development of expert systems, image feature recognition, real-time data access, and database systems. Near-real-time data assimilation is also required for initializing and updating forecast models.

Space weather services depend on data collection and processing in the same way that tropospheric weather services do. Data need to be collected from a large number of sensors strategically placed on Earth's surface, in Earth orbit, and in interplanetary space. The forecast centers need computer and communication systems that can rapidly process and analyze large volumes of observational data; run complex models in real time; display and manipulate imagery; derive, generate, and disseminate useful products; and facilitate data sharing and backup responsibilities. Acquisition of new data sets and development of advanced models, with complex calculations, will require greatly enhanced computer systems at the core of space weather services. The forecast centers must replace and continually upgrade both hardware and software to deal with the growing computational and communications needs. The CCMC and RPCs provide a foundation for these capabilities, and the SEC is building its National Space Weather Information System to address these needs.

5.3.1   National Space Weather Information System

SEC began development and implementation of three new general purpose information systems in FY98 and FY99, including an information dissemination system (IDS), a data display system (DDS), and a data simulation system (DSS). The goal of this effort is reduced maintenance, increased system and network security, expanded operability, better management and control of user access, maximized fault tolerance, and a reduced number of unique interfaces required for each application. The IDS will be an open and extensible system that provides a common interface to SEC data stores (real-time and historical data, model output, data simulations, etc.) for local and remote systems and through a local area network, a dedicated wide area network, and the Internet. The DDS will be a separate but related development project to provide a single system which will consolidate and replace SEC present data display systems and web-based display systems. The DSS will operate through the IDS to provide simulated real-time data for SEC and NASA/Space Radiation Analysis Group (SRAG) applications including the DDS.

5.3.2　Data and Information Management Policies and Interaction with Vendors

The DOD and NOAA SEC space weather operations centers each have different customers with different requirements. Beyond providing alerts and warnings for the protection of life and property, the SEC is additionally bound by legal proscriptions defining the extent to which they may tailor products versus providing generic data to value added resellers in the private sector. In the DOD, finely tailored products are the rule rather than the exception but in some cases more generic data may be provided for use in distributed decision support systems across the department. This information is communicated through common user channels to the maximum extent possible. The SEC has developed a set of policies to guide its data and information management, particularly in light of commercial vendors. Additional information is available at the SEC's web site at http://www.sec.noaa.gov/.

5.3.3　Archiving and Analyzing Space Weather Data

Climatological studies and products must also be improved to satisfy the needs of planners and engineers to know the range of conditions their systems may encounter and the probabilities of those conditions. However, before improved climatological studies may be conducted, a more usable archive of space weather data is needed and a database of space weather impacts is also needed to better correlate space weather events with their effects.

Space weather data are not all archived and what is archived is distributed across a variety of locations and agencies and in both research and operational forms. The NSWP plans to develop a structured approach that will unify the various sources of data and establish a set of common standards for metadata (data about the data). Critical needs in this area are a system of quality control for the data and an effective method of archiving checked data so that they are readily accessible to the research community. In addition to quality controlled archives, it may be necessary to provide at least some "raw data sets" for use in testing models intended for operational use, providing the opportunity to test the model with sometimes noisy, sometimes incomplete input.

Some progress has been made in developing an impacts database, but national security and proprietary commercial concerns will continue to mitigate against collecting this information. The NSWP will continue to seek innovative ways to address these concerns and expand on this database.

5.4 Summary

Although the transition of research into operational capability continues to be a problem across the spectrum of scientific activity, significant progress has been made in bridging this chasm in space weather. The development of the Community Coordinated Modeling

Center (CCMC) paves the way to build research models already structured for easier transition and the Rapid Prototyping Centers (RPCs) provide the vehicle to test models in an operational setting, validate them, and seamlessly integrate them into daily operations.

The NSWP must continue to encourage interagency and international collaboration to obtain research observations for operational use wherever possible and to expedite the development and deployment of the necessary operational observing systems.

The NSWP will continue to promote and coordinate development of improved standards for data collection, formatting, communication, and management. This improvement will facilitate processing and use of the increasing volume of observations, model output, and products within and between operational centers and customers.

CHAPTER 6

EDUCATION AND OUTREACH

The National Space Weather Program (NSWP) and a generally increasing interest in space sciences have created and benefited from educational activities aimed at grade school children, college students, users of space environment data, forecasters, and the general public.

NSF has made one dedicated award for space weather education and outreach activities under the NSWP. This award, to the Space Science Institute, supported the development of educational materials. Though this has been the only outreach product supported by the NSWP, many educational and outreach activities have been supported in connection with on-going programs, and all have demonstrated tremendous success. Examples of these projects are NASA/ISTP outreach, the Windows to the Universe project at the University of Michigan, Creating the Public Connection at Rice University, and the education and outreach efforts conducted by the Space Environment Center in Boulder, Colorado.

Space weather observing and forecasting is growing as an exciting and appealing area of study for a broad spectrum of students. The NSWP will continue to develop efforts that will use this natural interest in space to further national science and mathematics education goals. Those goals, as enunciated in documents such as Benchmarks for Science Literacy (developed by the American Association for the Advancement of Science), National Science Education Standards (developed by the National Academy of Sciences), and the National Council of Teachers of Mathematics Standards, stress the need for students to acquire "process skills." These include the ability to observe and measure, to manipulate those data in meaningful and quantitative ways, to draw conclusions from such investigations, and to communicate those conclusions to others effectively. These documents call for closely tying science to technological and social issues so that students can see the relevance of science in their lives. The NSWP will continue to contribute to achieving these goals by supporting both formal and informal science education.

6.1 Formal Education

The Space Science Institute, NOAA Space Environment Center (SEC), NASA, and others are already providing educational materials at the K-12 level through a variety of resource materials and the growing number of Internet web sites related to space weather. For example, the SEC produced the "Solar Physics and Terrestrial Effects" curriculum guide for teachers of grades 7 through 12. It includes text on space weather and its effects, complete with problems and questions, and it includes eight different activities ranging from building a spectroscope, to collecting and analyzing data, to a game exploring the radiation hazards in space. SEC staff members routinely speak at schools, judge science fairs, and participate in teacher training, reaching more than 700 students and teachers in 1997 and 1998 alone. The Space Science Institute partners with educators to create curriculum guides and multi-media materials for use in the classroom.

The interest students have in space means that the materials and programs to support formal science education in schools developed as part of the NSWP will have widespread impact. The program should continue to support the development of instructional modules on problems related to space weather, such as flares, coronal mass ejections (CMEs), the solar wind, geomagnetic storms, and the effect of solar activity on Earth's climate. Although much of the effort to date has focused on grades K-12, modules should be designed to teach science to a general college audience by exposing them to the field's current research. The module on geomagnetic storms, for instance, might be incorporated into an Earth sciences course that explores the near-Earth space environment. Such a module could be designed to demonstrate the importance of the magnetosphere-ionosphere coupling by briefly exploring current magnetospheric modeling efforts and showing how critically the models depend on the way magnetospheric currents enclose the ionosphere.

An example of the manner in which research and education can be integrated is the involvement of students in auroral observations. Using visual scanning and some recording photometers, the students would report on their auroral observations and would analyze the photometer data and send it back to a central site. Similarly, schools could be provided with digital magnetometers for measuring the magnetic field. Students would be instructed on methods to analyze the data for waves and variations and send the data to a central site for archiving. The SEC has begun this project but is seeking additional funding to place magnetometer kits in schools around the country.

6.2 Informal Education

In addition to formal science education, informal education plays an important role in the NSWP. The National Science Foundation (NSF) and other Federal agencies stress that to foster a scientifically literate citizenry, science education needs to extend beyond formal education into centers of informal learning such as museums and science centers. These institutions need to reach people of all ages, interests, and backgrounds. Museum programs and interactive exhibits on space weather themes will be developed in close collaboration with the science museum community, including forging a long-term relationship with the Association of Science-Technology Centers.

The explosion of the World Wide Web in the last five years has carried to the public a wide array of space environment information and more and more articles are appearing in the popular literature. The SEC's Education and Outreach web page includes short papers on selected space weather topics, a primer on the space environment, and a glossary of space weather terms. They also produced a publication entitled "*Web Activities Using Scientific Data*" that includes space weather activities for children. More than 500 people have toured the SEC facilities to gain a better understanding of space weather, its impacts, and how the Nation responds. In addition, the SEC fields several television interviews as well as about 20 print media interactions each year. Sixteen groups have visited SEC between May 1999 and May 2000 to film space weather activities in the operations center.

In a novel approach to public outreach, the SEC cooperated with the U-Haul Corporation to place a space environment graphic design on about 200 of U-Haul's rental trucks. The design highlights the SEC, Boulder, and Colorado in U-Haul's "Venture Across America" series. Other complementary efforts include television programs on The Discovery Channel™ and the development of an IMAX™ film on solar maximum.

In the aviation arena, the Federal Aviation Administration (FAA) distributed a pamphlet entitled "*Why You Should Be Interested in Space Weather*" at the 1999 Experimental Aircraft Association annual convention and fly-in in Oshkosh, Wisconsin. The pamphlet provided a general overview of space weather and its effects with emphasis on impacts on communications and the accuracy and availability of the Global Positioning System (GPS).

6.3 Educational Programs for Space Scientists

Simultaneously, programs will be developed to provide space scientists with the latest information concerning science education and to train interested space scientists to work effectively with students, teachers, and science museums. Efforts by Dr. Bruce Alberts, President of the National Academy of Sciences, and others have shown that the informed participation of scientists can contribute significantly to science education reform. Thus the educational programs and materials developed by the NSWP will be informed by the best current thinking in science education. Within the American Geophysical Union, the Space Physics and Aeronomy section's Education and Public Information Committees have both worked to mobilize the space physics community in support of education efforts tied to the NSWP.

The SEC's 1999 Space Weather Week brought together the operator and research communities for each to gain a better appreciation of the other's activities, giving the research community an opportunity to better understand the needs of the user and operator community. This exchange is expected to continue in future SEC Space Weather Week events.

6.4 Educational Programs for Operations Personnel and Space Weather Customers

Education in space weather has historically focused on increasing and spreading knowledge within the scientific community, but less on educating users, operators, and other customers of systems that are affected by space weather. Although considerable progress has been made, better balance in this area is necessary to meet the NSWP goals, which can only be accomplished with advocacy and stated needs from educated users, operators, and policy-makers from the private, non-Department of Defense (DOD) government, and military communities.

DOD has a variety of programs to train space weather analysts, forecasters, officers, and advanced-degree officers. Advanced courses are provided to personnel assigned to duty directly related to space weather observing and forecasting. All newly commissioned Air Force weather officers receive introductory space weather training, and some officers are selected each year to attend graduate space weather programs. Space weather training has also been added to airman and noncommissioned officer weather skill courses that have traditionally focused on tropospheric weather, adding significantly to the number of military personnel with space weather knowledge. In addition to initial skills training, the Air Force Weather Agency has published "*The Space Environment: An Air Force Weather Informational Guide*" to provide study material for Air Force Weather personnel who did not receive formal training in the space environment and its impacts.

Training is also being provided to DOD operators (including those who "drive and track satellites") and to multiple other agencies (customers) conducting operations affected by space weather. Briefings are provided to senior military leadership on the impacts of space weather, and space weather fundamentals and impacts have been added to space operator, space tactics, and other formal military courses. The Air Force Directorate of Weather has also produced an automated, narrated space weather presentation entitled "Space Weather and the Solar Maximum: Impact to Military Operations and What You Can Do About It." Space weather effects have recently been incorporated into several large military exercises and modeling and simulation of the natural environment continues to expand rapidly, including the incorporation of space weather effects in wargaming scenarios.

At the Space Environment Center (SEC) formal training courses are provided for personnel assigned as space weather analysts and forecasters. In addition, the synergy between the co-located operational analysts/forecasters and the scientists of the SEC Research and Development Division affords a unique opportunity. Research staff provide analysts and forecasters with improved understanding of the applicable physics, especially when new or unusual conditions are identified in the space environment, and in return, operations staff offer insights to the scientists on the practical applications of space weather science.

Civilian customers of space weather information require education, often at short intervals, because of constant turnover of industrial staff. On the other hand, some customers are more knowledgeable about the space weather effects on their own systems because of their life-long involvement. This inhomogeneous mix of customers makes the civilian user education process a multi-tiered problem, one that needs immediate and constant attention. NOAA's Space Environment Center has included space weather tutorials as part of its Space Weather Week and User Conferences in both 1998 and 1999. These sessions were well attended by a range of users and providers and the SEC produced a 90-minute "*Introduction to Space Weather*" videotape based on the 1998 training session. The SEC also publishes its "SEC User Notes" monthly as part of its Customer Focus Group efforts to communicate with their customers.

To enhance the education of current and emerging space weather customers, the SEC has maintained an active relationship with the public media and non-specialized science programs. They launched a campaign to make space weather more understandable and began creating special advisories in plain English to be distributed when major events occurred. Shortly thereafter, SEC developed and announced a major new method of providing space weather information through the use of space weather scales that classify events on a five-level scale that can be applied across three major types of storms. The scales also include the climatology, expected terrestrial effects, and measurement basis for each scale. Though developed for the general public, the scales have been widely accepted as a basis for space weather services by dedicated users and already have extended the use of space weather information to new applications.

The SEC also collaborated with the American Association for the Advancement of Science (AAAS) for a special session during their February 2000 annual meeting entitled "Space Weather and Things that go Bump in the Day." The session featured invited talks on space weather and its effects. A press conference associated with the meeting was one of three major press conferences about space weather organized by NOAA in the last two years.

Along with the recent growth in attention to space weather in the popular literature, many citizen groups, school classes, and science fair aspirants have shown an interest in the field, all of which clearly serve the NSWP. Although the training of analysts/forecasters and the expansion of knowledge through research are fundamental to the program, further outreach to other audiences (to gain advocacy, funding, and requirements) remains a crucial step to success. More information on SEC's public education and outreach efforts is available at their web site at http://www.sec.noaa.gov/.

CHAPTER 7

PROGRAM MANAGEMENT

The National Space Weather Program (NSWP) has been implemented by scientists, engineers, and technicians in government, academia, and industry. The program builds on existing capabilities and establishes an aggressive, coordinated process to set national priorities, focus agency efforts, and leverage resources to gain the biggest return. Organization and planning of the program requires a structure guaranteeing effective feedback and communication between the various communities involved. Appendix D contains points of contact and sources of information to facilitate this interaction.

7.1 Management Structure

The management structure for NSWP, organized within the OFCM includes the National Space Weather Program Council (NSWPC) and the Committee for Space Weather (CSW).

7.1.1   National Space Weather Program Council (NSWPC)

NSWPC is a multi-agency group designed to provide oversight and direction to the integrated process of setting national priorities, focusing agency efforts, and leveraging existing resources. It was established with the approval of the Federal Committee for Meteorological Services and Supporting Research (FCMSSR) in December 1994. NSWPC establishes policy, coordinates interagency efforts, and approves interagency agreements developed within the scope of the program. It also defines and coordinates the implementation of the NSWP. The NSWPC ensures that common needs are met and the interests of each agency are addressed. Member agencies retain responsibility for planning, programming, and budgeting their own resources to meet agency obligations to the NSWP.

NSWPC consists of designated representatives from Federal agencies involved in space weather activities. The representatives are the official spokespersons for their agencies on matters such as program scope, requirements, and resource commitments. Agencies involved are NSF, NASA, and the Departments of Commerce, Defense, Transportation, Energy, and the Interior.

7.1.2   Committee for Space Weather (CSW)

The CSW is aligned under the NSWPC and functions as a steering group responsible for tracking NSWP progress, identifying problems that threaten to delay or interrupt the program, and recommending corrective actions to the Program Council. Like the Program Council, the CSW is a multi-agency organization and is also composed of representatives from NSF, NASA, and the Departments of Commerce, Defense, Transportation, Energy, and the Interior.

7.2 Relationship Between NASA's Living With a Star Program and the NSWP

Living with a Star (LWS) is an exciting new NASA initiative that will provide major contributions to the National Space Weather Program (see Appendix B). LWS will accelerate the deployment of the present Solar Terrestrial Probe series and establish a new set of space weather research satellites (the Space Weather Research Network). LWS will also provide significant funding enhancements for space weather data analysis, theory and modeling. LWS provides key observational assets for testing space weather models, conducting basic space weather research and prototyping new operational platforms and instrumentation.

To ensure maximum benefit to the NSWP, LWS activities will be coordinated with other NSWP activities through NASA membership on the Committee for Space Weather (CSW) and by CSW members' participation in LWS planning. Furthermore, to ensure needed coordination of the basic research components of the NSWP, NASA and NSF intend to establish a collaborative research, data analysis, and modeling effort in space weather by combining efforts in the annual NSF/DOD NSWP research competition with the NASA LWS research announcements.

7.3 National Security Space Architect (NSSA)

The National Security Space Architect (NSSA), under the Office of the Assistant Secretary of Defense, is currently completing a two-year study of the Department of Defense's requirements for space weather services for the next 15 to 25 years. On December 4, 1997, the Space Weather Architecture Study Terms of Reference directed the NSSA to lead an integrated Space Weather Architecture Study with the DOD, NASA, NOAA, and other government agencies. Accordingly, the NSSA formed a Space Weather Architecture Development Team (ADT) composed of representatives from major stakeholders and conducted an architecture study to develop architecture alternatives.

The Space Weather Architecture Study was conducted in two phases. Phase I determined that an architecture study was warranted and gathered the information necessary to conduct it. Phase II developed and analyzed architecture alternatives and generated Space Weather architecture findings and recommendations.

After the study was completed, the National Security Space-Senior Steering Group (SSG) endorsed an Architecture Guidance Memorandum that identified the Assistant Secretary of Defense for Command, Control, Communications, and Intelligence (C3I), in coordination with NOAA, as the overall agency responsible for overseeing a Space Weather Transition Team, composed of key space weather stakeholders. A Space Weather Transition Team was organized to develop a plan to provide guidance on implementing the approved recommendations. The recommendations are summarized in Appendix C.

The NSSA study represents a detailed examination of space weather requirements over the next two and a half decades. Although organized by the DOD, the involvement of NOAA, NASA, NSF, FAA, and other non-DOD agencies ensured that the recommendations reflect the needs and roles of the commercial space weather customers, as well as researchers in government, academia, and industry. Because of its critical role in coordinating interagency efforts in space weather, the Committee for Space Weather will provide oversight to ensure continued pursuit of the NSSA recommended activities. The NSSA recommendations consist of a set of actions and activities that must be accomplished to achieve the desired architecture. The CSW will monitor these activities and update them as necessary. Because of the emphasis on defense requirements, some of the activities may differ slightly from priorities set earlier by the NSWP. Members of the CSW will coordinate to resolve these differences. Adjustments to the NSSA recommendations will be made only with full concurrence of the DOD membership on the CSW. The NSSA recommendations represent an aggressive and ambitious program to achieve improvements in space weather capabilities as quickly and as cost effectively as possible. This detailed study will provide an effective means for planning and organizing activities and for tracking progress in space weather.

7.4 Coordination with the Research Community

Active involvement from all space weather stakeholders has been a high priority since the inception of the program. Formal and informal mechanisms have been used to maintain effective coordination among space weather researchers, instrument developers, data providers, operational forecasters, and customers. On the research side, space weather information is exchanged at semiannual meetings of the American Geophysical Union and annual workshops convened by the CEDAR, GEM, and SHINE communities. Over the past three years, all of these venues have featured special sessions on Space Weather.

To facilitate communication among researchers, the operational community, and space weather customers, the Space Environment Center has established Space Weather Week, the first of which occurred in April 1999. The first part of this week is focused on the transition of research to operations, meetings which were initially held separately in January of 1997 and 1998. Space Weather Users Meetings had previously been held every three years, but have now been combined with the research to operations meetings to form the second half of Space Weather Week. Together, these meetings provide an excellent forum to bring researchers together with both the operations and user communities.

Aggressive research programs supporting space weather goals continue to be supported by NSF, NASA, NOAA, and the DOD. Many areas of research in space and plasma physics directly support program objectives by advancing knowledge in fundamental scientific areas. However, early in the program it was recognized that rapid progress could be made only by implementing a more targeted research program. Toward this end, NSF, AFOSR, and Office of Naval Research (ONR) contributed to funding competitively-selected research proposals in key areas. These competitions were held in 1996, 1997, 1999, and 2000. The program announcements included a description of the areas of scientific emphasis to fill gaps in our existing knowledge and predictive capabilities. The proposals were selected by panels who evaluated them on the merit of the research and their potential to contribute to space weather goals. As indicated in the description of research in Chapter 3, these awards included many innovative approaches to space weather model development and predictive capabilities, as well as more basic research aimed at improving our understanding of space weather phenomena.

In the future, the Committee on Solar Terrestrial Research (CSTR) of the National Academies will periodically review past progress and make recommendations to the CSW on areas of emphasis for future space weather proposal competitions. On the basis of those recommendations, as well as the comments and insights of the research community, CSW will formulate the updated space weather announcement of opportunity.

7.5 Coordination with the User Community

Equally important for effective progress in space weather goals is to establish a good interface with space weather customers. Interaction between space weather customers and operational forecasters has previously taken place at the Space Environment Center in Boulder during the Space Weather User Conferences held every three years. In 1998, meeting attendees favored meeting annually, at least during the solar maximum period, and also endorsed merging the meeting with the Space Weather Research to Operations workshops previously held in January in Boulder. The two meetings were held consecutively for the first time in April 1999 in an event referred to as Space Weather Week. This successful format for encouraging feedback between the scientific community and space weather customers will continue on a yearly basis for the next several years.

In addition to the Users Conference, two other more specialized workshops were conducted. The first was a workshop on Geomagnetically Induced Currents held at the

Electric Power Research Institute (EPRI) headquarters in Washington, D. C., during October 1996. The second was a workshop on Space Weather Effects on Navigation and Communication Systems which was held at COMSAT headquarters in Bethesda, MD, in 1997.

Identifying the requirements of the satellite industry has been difficult due to the highly competitive nature of the industry, the complexities associated with insurance coverage, and the legal aspects of satellite communications. To initiate discussions with satellite industry representatives, NSF made an award to Sterling Software whose representatives, armed with nondisclosure agreements, interviewed top industry executives. The DOD has also initiated efforts with each of its services to improve and expand the documentation of space weather impacts on military operations.

7.6 Non-Federal Involvement

NSWP is a cooperative effort within several agencies of the Federal government. However, achieving the goals of the program requires the participation of a variety of entities outside the Federal government, including universities, research institutes, laboratories, and businesses. Indeed, the role of the Federal government, through OFCM's CSW and cooperating agencies, is largely one of management and coordination. Several areas provide opportunities for participation by non-Federal entities.

Requirements. Requirements for space weather support should be forwarded to the appropriate agency responsible for providing support. DOD agencies should request support from the Air Force Weather Agency. Other government and private agencies should request support directly from the Space Environment Center (SEC). Support requirements that exceed the current state of the art should be stated as early as possible. See Appendix D for addresses.

Research Opportunities. In addition to the NSWP targeted research opportunities mentioned in Section 7.3, other research opportunities are expected to be available through the Federal agencies that support the program. Although these opportunities will not be explicitly tied to the NSWP, they will support the broad goals of the program and, in many cases, its specific objectives. From DOD, opportunities will be available through the Air Force Office of Scientific Research, the Office of Naval Research, and for developmental research, the Defense Modeling and Simulation Office (DMSO) through the Air and Space Natural Environment Modeling and Simulation Executive Agent. DOC, through NOAA's Office of Oceanic and Atmospheric Research, will offer opportunities as will NASA's Office of Space Sciences. On a more limited basis, DOE will offer some opportunities through its Solar Terrestrial Research Program within the Office of Basic Energy Science.

Rapid Prototyping. A Rapid Prototyping Center (RPC) is already in place at the SEC and another RPC focusing on military requirements is in development through the Air Force's Space and Missile Systems Center (SMC), Air Force Space Command, and the Air Force

Weather Agency. The Community Coordinated Modeling Center (CCMC) is an additional link in the chain as a feeder process to the RPCs, bringing research models into the operational sphere. Procedures to nominate appropriate models to these centers have been established and more information is available from the centers themselves. See Appendix D for instructions on contacting these organizations.

Producing Tailored Products. Not all space weather support requirements will be met by products issued by the SEC. Some civilian operators will require very detailed forecasts for specific weather elements at specific places and times. The opportunity exists to access SEC products, tailor them to customers' specific requirements, and disseminate them to operators as a business opportunity. Although this type of effort receives cooperation as appropriate from Federal agencies, it should not expect to receive Federal support.

7.7 Agency Roles and Responsibilities in the National Space Weather Program

DOC (represented by NOAA), DOD, DOT (represented by the Federal Aviation Administration), NASA, DOI, DOE, and NSF recognize common interests in space weather observing and forecasting. Aware of the need for prudent employment of available resources and the avoidance of duplication in providing these services and support for agency mission responsibilities, the cooperating agencies have sought to satisfy the need for a common service program under the NSWP. This section provides information on how the participating agencies contribute to the program today and in the future. The general information on each agency provided in the following subsections is further detailed in Table 7-1, which indicates which space weather domains (from Table 2-2) the agencies address, and in which areas they intend to focus their efforts.

NOAA and the Air Force have separate, distinct, statutory roles in providing space weather observations, forecast and warning services, and data archival to the civil sector, DOD, and other Federal agencies. NOAA, through the SEC, provides centralized space weather support to non-DOD government users (e.g., NASA) and to the general public. The United States Air Force (USAF), through the 55th Space Weather Squadron (55 SWXS), provides unique and sometimes classified support to all DOD users. To avoid duplication, the two agencies share responsibilities to produce certain space weather databases, warnings, and forecast products. Both agencies also support space weather research.

The 55 SWXS and the SEC provide cooperative support and backup for each other in accordance with existing agreements. USAF assigns personnel to Operating Location A, 55 SWXS, collocated with the SEC, to assist in the operations of SEC and to participate in activities of mutual interest and benefit to the USAF and NOAA.

NASA and NSF play key roles in advancing operational space weather support through research. Both agencies deploy systems that collect data to support research focused on

Table 7-1. Agency Participation Matrix

	Physical Understanding	Model Development	Observing Systems	Technology Transfer
Solar coronal mass ejections	1,2,3,6	1,2,5,6	1,2,3,5	1,2
Solar activity/flares	1,2,6	2,5,6	1,2,5	2,7
Solar and galactic energetic particles	1,2,3,6	2,5,6	1,2,3,5	1,2
Solar UV/EUV/soft x-rays	1,2,6	5,6	1,2,5	1,2,7
Solar radio noise	1,6	5,6	1,2,5	1,2
Solar wind	1,2,3,6	1,2,5,6	1,2,3,5	1,2
Magnetospheric particles and fields	1,2,3,4,5,6	1,2,3,4,5,6	1,2,3,5	1,2,3
Geomagnetic disturbances	1,2,3,4,5,6	1,2,3,4,5,6	2,3,4,5	1,2,3
Radiation belts	1,2,3,5,6	1,2,3,5,6	1,2,3,5	2,3
Aurora	1,2,3,4,5,6	1,2,4,5,6	1,2,3,5	1,2
Ionospheric properties	1,2,3,4,5,6	1,2,4,5,6	2,3,5	2,7
Ionospheric electric field	1,2,4,5,6	1,2,4,5,6	1,2,3,5	2,7
Ionospheric disturbances	1,2,3,4,5,6	1,2,4,5,6	2,3,4,5	2,7
Ionospheric scintillations	1,2,5,6	1,2,5,6	2,5	2,7
Neutral atmosphere (thermosphere and mesosphere)	1,2,5,6	1,2,5,6	2,5	1,2

Organization codes: 1=DOC, 2=DOD, 3=DOE, 4=DOI, 5=NASA, 6=NSF, 7=DOT improving our understanding of space weather processes. They also manage much of that research.

DOI and DOE participate by collecting data that, while supporting their missions, contribute to the operational space weather database. They also support limited research related to those data.

DOT, through the Federal Aviation Administration, participates as a regulator of the commercial space industry and as an implementor and operator of advanced Global Positioning System (GPS)-based systems for air navigation. Space weather's adverse effects on manned space flight, high altitude aircraft operations, and the availability and accuracy of GPS-based navigation make understanding and mitigating these impacts an imperative for the FAA.

7.7.1   Department of Commerce (DOC)

DOC's NOAA is responsible for monitoring and forecasting the near-Earth space environment for nonmilitary applications. NOAA's programs support governmental, commercial, educational, and scientific communities. Activities focus on satellite instrumentation, data assimilation, environmental forecasting, and research and numerical modeling. NOAA and DOD cooperate on programs of mutual interest.

Currently, NOAA operates space environment instruments on Geostationary Operational Environmental Satellites (GOES), and polar-orbiting satellites to monitor solar emissions and in situ plasma fluxes. In the future, NOAA will operate the joint DOC-DOD National Polar-Orbiting Operational Environmental Satellite System (NPOESS). In addition, NOAA has proposed a solar wind monitoring program that would support the NSWP.

NOAA's SEC and National Geophysical Data Center (NGDC) receive, process, analyze, and assimilate space weather data collected by worldwide networks of satellite and ground-based instruments. SEC is responsible for real-time and operational data. NGDC is responsible for the national and World Data Center archives. Future data activities will focus on greater spatial coverage of relevant parameters from national and international partners.

Nowcasts and forecasts are routinely prepared and distributed by SEC. Numerical analyses and model simulations are conducted by SEC and NGDC. In support of the NSWP, SEC has been working to test, evaluate and incorporate superior algorithms to forecast the space environment.

Research and modeling activities at SEC and NGDC include analysis of in situ measurements and development of numerical models conducted by government scientists and international visitors. In support of the NSWP, NOAA through its Rapid Prototyping

Center tests and evaluates physical models developed by academia, government, and industry under routine, near-real-time conditions.

7.7.2   Department of Defense (DOD)

7.7.2.1   *Support for Observing, Forecasting, Modeling, and Research*

DOD will continue to support observing, forecasting, modeling, and research efforts supporting operational assets in the near-Earth space. Through the Defense Modeling and Simulation Office (DMSO), DOD supports development of authoritative representation of the near-Earth space environment for use in joint service modeling and simulation programs. Through 55 SWXS, DOD will continue to monitor data from various ground sites and space-based observation platforms to provide warning, observing, and forecasting support for both military and civilian assets in conjunction with SEC.

In support of the NSWP, the Air Force Research Laboratory (AFRL) and various contracting agencies develop modeling techniques for use at 55 SWXS. AFRL will continue to be the focal point for space weather models for DOD and models from outside agencies will be validated and transitioned via the Rapid Prototyping Center with the help of AFRL.

DOD will support research to improve the understanding of space weather phenomena, particularly in the near-Earth regions. The department recognizes that knowledge of solar and interplanetary phenomena is critical to forecasting in the magnetosphere and ionosphere, and it continues to advocate research in those areas as well. Efforts to develop sensors and spacecraft to measure the space environment will be leveraged with other agencies to build and deploy effective platforms. Data from these systems will then be assimilated into the operational models and archived for climatological studies.

7.7.2.2   *National Security Space Architect*

The Office of the National Security Space Architect (NSSA) will track progress made toward the space weather architecture and recommendations resulting from their architecture study conducted between December 1997 and September 1999. As periodic reviews of the NSWP occur, the NSSA, Program Council, and CSW will coordinate actions and may enlist NSSA's assistance for a detailed review to match breakthroughs in technology or changes in requirements. The NSSA will be represented in this process via the DOD members on the NSWPC and CSW and will have full rights to identify issues or request an update to the architecture.

7.7.2.3   *Restructuring DOD Space Weather Support*

To improve space weather support to the DOD and National Programs, the Air Force is restructuring its space weather operations. This action directs an end-to-end restructuring of organizational and operational responsibilities for the space weather mission support area within the DOD. It will integrate terrestrial and space weather services within the Air Force, leverage Air Force Weather capabilities to improve the space weather mission area, and retain strong leadership by Air Force Space Command (AFSPC) for program acquisition and modernization.

Under this restructuring, the Air Force Weather Agency (AFWA) will provide space environmental information to all DOD and National Program users. As part of this restructuring, the 55th Space Weather Squadron (55 SWXS) was realigned under AFWA on 1 October 1999. In this way, HQ USAF/XOW through AFWA, and the Air Force major commands through their operational- and tactical-level units, share responsibility for providing and applying space weather information for military operations.

The Air Force is in the process of establishing an improved space weather center capability within the AFWA infrastructure at Offutt AFB, Nebraska. As this capability becomes operational, the responsibility for providing space weather support will transfer from the 55 SWXS at Schriever AFB, Colorado, to the new AFWA Space Weather Operations Center. This transition will occur over the period from the summer of 2000 through the first quarter of Fiscal Year 2003.

This restructuring improves the Air Force's space weather support organization in order to enhance the mission effectiveness of DOD and National Program operations and planning. The basic plan has been developed to create the most effective organizational structure and technical capability to provide mission-tailored space weather products. Fused with terrestrial weather information, these products will yield an integrated, "mud to sun" analysis and forecast of the environment through which DOD and National Program missions are conducted.

This effort is intended to improve space weather mission support in several ways. These include:
- Realignment of space weather information providers to better meet strategic-, operational- and tactical-level mission needs
- Improved timeliness, accuracy, and relevance of space weather information to better focus on national defense needs
- Improved integration of space weather providers into their customers' operations and planning, which will foster better application of force enhancement information
- More rapid integration of emerging space weather technologies to meet operational requirements
- Improved training to provide more fully qualified space weather providers and users

- Improved structure for the collection and validation of space weather requirements

A major benefit of this plan is to spread space weather expertise throughout Air Force Weather forces, raising awareness of the effects of space weather on DOD and National Program operations and enhancing mission accomplishment. In summary, this restructuring effort addresses those areas of space weather services which can be improved to meet current and future national defense requirements.

7.7.3   National Science Foundation (NSF)

NSF supports and will continue to support basic research in solar-terrestrial sciences, including the Sun, solar wind, magnetosphere, ionosphere, and thermosphere. NSF supports theoretical and observational research with the goals of increasing fundamental understanding of space environment processes and improving space weather predictive capability. The research includes the development and operation of ground-based space environment monitoring instrumentation; the development of ionospheric, thermospheric, and magnetospheric specification models; and the analysis of post-event databases. The worldwide array of NSF-sponsored instruments, observatories, and facilities will continue to provide vital ground-based measurements in coordination with space missions sponsored by other agencies.

Research areas of emphasis are (1) solar region evolution and eruptive events, (2) interplanetary transport, (3) magnetospheric physics and dynamics, (4) ionospheric physics and dynamics, and (5) upper atmosphere physics and dynamics. Knowledge of the processes that are fundamental to each of these areas will be enhanced by a multi-disciplinary approach to investigating the basic mechanisms through which these areas interact.

7.7.4   National Aeronautics and Space Administration (NASA)

NASA will continue its traditional role of research in the physics of the solar-terrestrial system. This research program is carried out under the theme "Sun-Earth Connection" and now "Living with a Star," both of which seek to explore and understand as one system the dynamics of the Sun and its interactions with Earth and other planetary bodies and with the interstellar medium. Key questions addressed relevant to space weather include the following: What causes solar variability? How does the Sun and its variability affect Earth and other planetary space environments? The Sun-Earth Connection uses the solar system as a laboratory to understand basic plasma physical processes such as the acceleration of particles to high energies and generation of intense radiation belts or plasma enhancements, processes that can affect electronic and biological systems exposed to the space environment.

The ongoing and future space science flight programs of NASA and its partners are making, and will continue to make, critical contributions to space weather research. The International Solar Terrestrial Physics (ISTP) Program is providing new experimental and theoretical advances in solar-terrestrial physics. The ISTP missions (Geotail, WIND, POLAR, SOHO), the complementary missions (Yohkoh, Fast Auroral Snapshot (FAST), IMP-8, SAMPEX, ACE, TRACE, IMAGE), and TIMED will significantly advance the state of knowledge in solar-terrestrial physics. ACE provides the first 24-hour-per-day broadcast of real-time solar wind data used for space weather forecasting by the DOD and NOAA. NASA is a partner with USAF in the development of the Solar Mass Ejection Imager (SMEI). Future missions under consideration include a solar stereo mission that could provide images of coronal mass ejections directed toward the Earth and arrays of microsatellites to provide multi-point measurements in the magnetosphere.

NASA missions are designed under several guiding principles:
- To improve and advance empirical understanding of events and conditions in space
- To develop and use new technology
- To establish proof of concept and the value of new observational methods in space (e.g., energetic neutral particle imaging of the magnetosphere)
- To develop a database that determines the empirical nature of space weather conditions
- To observe, interpret, and understand the causes of and to predict the variable particle and electromagnetic radiations that emanate from the Sun and affect the space environment of Earth and other planets.

Much of the Sun-Earth Connection research grants program also contributes to developing basic principles and methods by which space weather may be understood and predicted. More information on "Sun-Earth Connection" and "Living with a Star" is provided in Appendix B.

7.7.5   Department of the Interior (DOI)

DOI participates in the NSWP through its United States Geological Survey (USGS), which operates a series of geomagnetic observatories and participates in worldwide collection and real-time exchange of geomagnetic data. It provides these data to USAF and NOAA operational centers for determination of geomagnetic indices to support warning and forecasting and to NGDC for archiving to support research. DOI also conducts research in geomagnetic and electrical fields, particularly in how they relate to the structure of Earth's core and mantle.

In the future, DOI plans to expand its network of geomagnetic observatories. In the near term, expansion will provide data from land areas from which data have not been available in the past. In the longer term, USGS plans to organize an effort to collect geomagnetic data over the broad ocean areas. Research conducted or supported by DOI will continue to be focused in areas where it is concentrated today, but will evolve to exploit the increase in the number of geomagnetic observations available and the changing distribution of spatial coverage.

7.7.6   Department of Energy (DOE)

DOE will continue its ongoing program to supply energetic particle and plasma sensors at geosynchronous and Global Positioning System (GPS) orbits and to support the analysis and distribution of those data in a timely manner. In addition, DOE should be considered as a candidate agency to provide similar environmental sensors for other future magnetospheric monitoring tasks within the NSWP.

DOE's Los Alamos National Laboratory has supplied plasma monitors for many NASA solar wind missions. Through Los Alamos, DOE is providing real-time solar wind data from the ACE spacecraft, as well as the expertise necessary to support ACE. DOE should also be considered as a candidate agency to provide sensors for future NSWP solar wind spacecraft to follow ACE.

Los Alamos carries out an extensive program of space physics research that is sponsored by both DOE and NASA and that includes data analysis and interpretation as well as space plasma theory and modeling. DOE will continue to support this activity as a contribution to the research and modeling components of NSWP activity.

7.7.7   Department of Transportation (DOT)

Within DOT, the Federal Aviation Administration (FAA) is responsible for regulating and promoting the U.S. commercial space transportation industry. It licenses the private sector launching of space payloads on expendable launch vehicles and commercial space launch facilities. In addition, it also sets insurance requirements for the protection of persons and property and ensures that space transportation activities comply with U.S. domestic and foreign policy. Low-cost, reliable access to space is the foundation on which many other commercial and strategic applications of space technology are based. The benefits and spin-offs from these technologies, in turn, touch almost every aspect of the ability of the United States to remain at the forefront of world technological advancement and economic prosperity.

An important DOT space application is to successfully field a Global Positioning System (GPS)-based capability to support en route, terminal, and precision approach operations for airports and helipads/heliports in the U.S. and offshore areas such as Canada and Mexico. Towards this end, the FAA is developing the Wide Area Augmentation System (WAAS), a geographically expansive augmentation to the basic GPS service designed to improve the accuracy, integrity, and availability of the basic GPS signals. The WAAS will improve basic GPS accuracy to approximately 7 meters vertically and horizontally, improve system availability through the use of geostationary communication satellites carrying transponders, and provide important integrity information about the entire GPS constellation. DOT is working with the International Civil Aviation Organization to foster acceptance of a single Global Navigation Satellite System integrating this capability with other satellite-based augmentations worldwide.

7.8 International Space Weather Efforts

7.8.1 International Space Weather Environment Service

The International Space Environment Service (ISES)--formerly the International URSIgram and World Days Service (IUWDS)-- is the organization through which the United States participates in international, real-time exchange of data and forecasts for the space environment. ISES consists of regional warning centers in major areas of the world. The warning centers serve their own regions by collecting data and exchanging it for data from other warning centers. Each warning center provides daily forecast advice to the World Warning Agency, operated by NOAA as a part of the SEC in Boulder. Each day, the Boulder center issues a consensus set of forecasts and summaries of activity back to the regional warning centers. Data collected in near real-time include geomagnetic and ionospheric observations as well as other solar-terrestrial data. The ISES data exchange program is currently evolving as various centers convert their data services to use of the Internet and the ISES plans that various regional centers can assume responsibility for some part of the effort of providing forecasts and alerts. ISES is also evolving as a vehicle for arranging tracking of satellites such as ACE. DOC will work with the regional warning centers to arrange for additional collection of data needed for the NSWP and for cooperation in implementing improved space weather services.

ISES organizes a series of international workshops to evaluate requirements and methods of improving solar-terrestrial predictions. These are held at approximately 5-year intervals and cover methods of observing and forecasting activity from the Sun through the interplanetary space and into the neutral atmosphere, ionosphere, and magnetosphere. The last workshop was held in Japan in 1996. The next one is expected in 2000 or 2001.

As a way of improving the relevance of the workshops, ISES is considering a proposal to conduct coordinated international campaigns to test improved prediction techniques, with the involvement of scientists who have developed the techniques and the end users (forecasters and customers). If the proposal is accepted, SEC's Space Weather Operations is planning to be involved as a central forecast center in the prediction campaigns in several ways. This includes the coordination of campaigns, the provision of collecting data and the actual execution of the campaigns from the perspective of forecasting, research, and development. These campaigns will provide a window for international participation, with U.S. participation through the NSWP.

7.8.2   ESA Space Weather Program

The European Space Agency (ESA) organized a Workshop on Space Weather during 11-13 November 1998 at ESTEC, Noordwijk, The Netherlands. One of the goals was to determine the current state of the field in different countries. The second goal was to put together a global picture concerning all scientific, technological, economic and environmental issues concerning space weather with the emphasis being placed on defining potential user requirements for European Space Weather Services. Future space weather workshops are being planned.

ESA also solicited a report on the state of the art in space weather modeling and on a proposed ESA strategy. This report was prepared by the Finnish Meteorological Institute and published in October of 1998. It evaluates the present space weather requirements and capabilities worldwide, with particular emphasis on European plans and capabilities.

Following this report, ESA issued a tender to develop a space weather program tailored to the needs of the agency, with a deadline of September 13, 1999. This announcement of opportunity specifically requests analyses of needs, cost benefits, and detailed mission scenarios of a European Space Weather Program. In particular, the objectives of the solicitation are the following:

- Investigate the benefits of a space weather program
- To provide secretarial management of a European space weather working team
- To establish a detailed rationale for a space weather program
- To establish detailed program contents, including a space segment, and a definition and prototyping of services to be provided
- To define the structures which need to be implemented by ESA and member states
- To produce a draft program proposal, project implementation plan, cost estimate, and risk analysis
- To develop a web-based data base

The ESA announcement of opportunity mentions explicitly collaborations and coordination with international space weather programs and efforts such as the National Space Weather Program. Therefore, opportunities for coordination exist in the short and intermediate term with the consortium selected to conduct the ESA program study. It is to be expected that the ESA study will benefit from experiences obtained in the US both in the analysis of relevance, and in the prioritization of US efforts. Furthermore, US and European efforts can and should be coordinated to avoid duplication of efforts and for resource sharing. Because of the expected long time duration of the expected ESA program, this opportunity for collaboration extends into the far future.

7.8.3   Space Weather Programs in Other Countries

The *Swedish Institute of Space Physics* (IRF) studies how solar magnetic activity can be modeled and predicted with intelligent hybrid systems (IHSs) using SOHO data. They use neural networks to study and predict satellite anomalies from the space weather state, radio communication conditions (indicated for example by the foF2), and geomagnetically induced currents and their effect on electric power systems and gas pipeline systems. Global magnetic field variations have been predicted by using geomagnetic activity indices.

In September 1996, IRF planned to build and launch a very small student satellite to be named Munin after one of the god Odin's ravens. The scientific objective of Munin is to collect data on auroral activity in both the Northern and Southern Hemispheres, such that a global picture of the current state of activity can be made available on-line. The data acquired by Munin will then serve as an input to the prediction of space weather. Student projects involving the processing and reduction of the data are envisioned. This satellite project will be used for technology development and in Space Engineering Education run by the Department of Space Physics of Umeå University. The data collected will be published on the Worldwide Web, free for all to use.

The role of the *Australian Space Forecast Center* (ASFC) is to monitor and forecast the solar-terrestrial environment, the region of space encompassing the Sun, the solar wind, the Earth's geomagnetic field and ionosphere. To fulfill this role, the ASFC receives data from a network of solar, geomagnetic and ionospheric observatories within the Australian region. It also exchanges data with similar organizations in other countries to provide a continuous flow of solar-terrestrial information.

In France, a forecast center is a regional center of the ISES (International Space Environment Service), dependent on the ICSU (International Council of Scientific Unions). It is located at the Paris-Meudon Observatory, in the Solar Department (DASOP), and the scientific service is known as COMPAS. It forms part of the *Laboratory of Solar and Heliospheric Physics* (CNRS). The user area covered officially by the Paris-Meudon Center is Western Europe (23% of the users are in France, 47% in the remainder of Western Europe). Further, 18% of the users are from Eastern Europe and 12% from the rest of the world. More than half of the users are scientific organizations (example: EISCAT), the others being related to telecommunications, monitoring of ionizing radiation, and especially to space activities. The Meudon Center cooperates with space agencies (mainly CNES and ESA, but it also provides forecasts to the Indian and the Canadian Space Agencies) for various applications. These applications include validation of scientific data, conditions in the terrestrial environment during satellite launches, causes of anomalies on board satellites, and dangerous re-entries of certain satellites of great mass (Skylab) or carrying nuclear generators (Cosmos 1402 and Cosmos 1900). However the most significant need at the European level remains the supply of data and forecasts for calculation of Earth observation satellite orbits—the series represented by SPOT (CNES), ERS (ESA) and Topex-Poséidon (CNES - NASA).

In Japan, the *Space Environment Information Service* of the Hiraiso Solar Terrestrial Research Center provides space weather services. These services include updates on solar and geomagnetic activity, providing near real-time data from the ACE spacecraft, and daily plots of high-energy particle fluxes measured by the Space Environment Monitor (SEM) on GMS-4, the Japanese geosynchronous meteorological satellite. These data, as well as ground-based geomagnetic field data, are available via web access. Similar web access is provided to ionospheric sounding data in Japan. The Hiraiso center also maintains the Space Environment Real-time Data Intercommunication Network (SERDIN). SERDIN is a core facility required for space weather forecasts and is designed to perform acquisition, analysis, and distribution of space environment data in an automated manner. A wide variety of solar and geophysical data are collected via either local area network at Hiraiso Center or the wide area network (domestic and overseas) links in near real-time.

In Canada, NRCan (Natural Resources Canada) provided forecast services while support for basic research on the space environment is obtained through NSERC (Natural Sciences and Engineering Research Council). The Canadian Space Agency (CSA) provides major funding of facilities for use by Canadian scientists. Because of Canada's high latitude location, its technological systems are considerably more vulnerable to adverse space weather. A long-term space plan has thus been proposed through the CSA to integrate the extensive remote sensing facilities into an enhanced, ground-based, Canadian super-array. A new national facility for data assimilation and modeling has also been proposed.

7.8.4   Coordination of International Space Weather Activities

The Scientific Committee for Solar-Terrestrial Physics (SCOSTEP), during its 9th Quadrennial Symposium at Uppsala in August 1997 helped organize a special evening session to consider international space weather issues. NSF participants gave a presentation on the US National Space Weather Program at that meeting. Similar reports were presented by representatives of a number of other countries which have space assets and desire to understand and forecast disturbances in near-Earth space that affect their operations. After the meeting, the group of several hundred participants voted to ask SCOSTEP to lead an oversight effort to provide a coordinated international space weather program. This will be done as part of S-RAMP, which stands for "STEP-Results, Applications, and Modeling Phase", a new program adopted by SCOSTEP for the period 1998-2002. S-RAMP is to be the follow-on program to STEP (Solar Terrestrial Energy Program). At the Uppsala meeting, a Space Weather Working Group of S-RAMP was set-up and the first meeting of this Space Weather Working Group was convened at the COSPAR General Assembly in Nagoya, Japan, in July 1998 to discuss future activities. A new International Space Weather Clearinghouse web site was established at the University of Michigan to provide a forum for this group's activities (URL: http://aoss.engin.umich.edu/intl_space_weather/sramp/).

The American Geophysical Union's fifth Western Pacific Geophysics Meeting was held in Taipei, Taiwan, in July 1998 after the COSPAR Meeting. Scientists representing the US, Japan, Taiwan, Korea, Canada, and Australia provided talks on their countries' national space weather programs. The S-RAMP Steering Committee met in Taipei and identified the especially active solar period of April-May 1998 as a "Special Study Interval". The Committee encourages scientists involved in all aspects of solar physics, interplanetary and magnetospheric physics, upper atmosphere and middle atmosphere physics, and space weather topics to concentrate on phenomena recorded by the extensive array of satellites and ground-based facilities operating at that time.

The Committee also approved the idea of organizing the First S-RAMP Space Weather Campaign for September of 1999. The objective of this campaign was to study the effects of space weather disturbances on the coupled magnetosphere-ionosphere-thermosphere system on a global scale, including the impacts on technological systems such as electric power grids, satellites, and ground- or space-based communication and navigation systems. In order to better understand the physical processes, as well as to provide a quantitative assessment of the effects on technological systems, the campaign also included observations from other global arrays of radio, optical and magnetic instruments. An effort was made to have efficient communications of activity alerts, information about observations, and electronic transfer of data and images worldwide. Special effort was and will continue to be made to see that scientists in developing countries have good access to data in a timely manner. The Space Physics and Aeronomy Research Collaboratory (SPARC) at the University of Michigan supported this campaign by providing real-time access to a wide range of space and ground-based data and model outputs. The results obtained during this campaign are expected to be discussed at the First S-RAMP Conference in Sapporo, Japan, in October 2000.

APPENDIX A

NATIONAL SPACE WEATHER PROGRAM RESEARCH

To achieve the goals of the National Space Weather Program (NSWP), we must improve our physical understanding of the space environment, improve empirical models, develop physics-based models that adequately specify and forecast that environment, and develop a suite of sensors to provide the observations necessary to drive those models. This appendix provides detailed information on how those objectives may be accomplished. For each of the three primary aspects of research--physical understanding, model development, and observations--the information is organized by space weather domain as presented in Table 2-2 or Figure 3-1. For reference, these domains are listed below.

- Coronal mass ejections
- Solar activity/flares
- Solar and galactic energetic particles
- Solar UV/EUV/soft x-rays
- Solar radio noise
- Solar wind
- Magnetospheric particles and fields
- Geomagnetic disturbances
- Radiation belts
- Aurora
- Ionospheric properties
- Ionospheric electric fields
- Ionospheric disturbances
- Ionospheric scintillations
- Neutral atmosphere (thermosphere and mesosphere)

Although it is convenient to deal with each of these domains separately, the coupling of the physical processes between the regions of the space environment represented in this list (solar/solar wind, magnetosphere, and ionosphere/thermosphere) must not be overlooked. *The goals of the National Space Weather Program can be achieved only when the representation of space weather is coupled into a seamless system, starting at the Sun and ending at the Earth.*

A.1   Physical Understanding

The greatest challenge in improving our ability to specify and forecast space weather is in understanding the physical processes that drive it. Until we improve this basic understanding, we cannot develop and deploy valid models. Nor can we adequately define our observational requirements. This section addresses the fundamental need for basic knowledge in the space weather domains listed above.

A.1.1   Coronal Mass Ejections (CMEs)

The research objective is to understand

- *the physics of the CME initiation process and the factors that determine CME size, shape, mass, speed, and internal field strength and topology.*
- *how to predict the above on the basis of planned observing systems.*
- *how to predict CME-caused solar wind disturbances and solar energetic particle events near Earth.*

The major non-recurrent geomagnetic storms are generally associated with the solar wind disturbances created by fast CMEs. In addition, most major solar energetic proton events observed near Earth seem to result from acceleration of some solar wind particles by the CME-driven interplanetary shock wave. Substantial advances in space weather forecasting can be made by learning more about the solar processes that initiate CMEs, and by learning how to predict their interplanetary consequences from observable signatures at the Sun and from in situ solar wind plasma, magnetic field, and energetic particle observations.

The slower CMEs eventually (far from the Sun) attain speeds comparable with that of the normal, slow solar wind, but do not generate significant solar wind disturbances, energetic particle events, or geomagnetic disturbances. On the other hand, the faster CMEs produce major solar wind disturbances as they overtake, compress, and accelerate the slower ambient solar wind ahead. Large CME-driven interplanetary disturbances are usually preceded by strong shocks that are effective accelerators of particles and sources of radio emissions. Strong magnetic fields are also commonly found in these disturbances both behind the leading shocks and within the CMEs themselves. These strong fields are primarily a result of compression caused by the interaction between CMEs and the ambient wind. When the compressed fields in the ambient wind and/or within the CMEs have substantial southward components, major geomagnetic storms result as they interact with Earth's magnetosphere.

A.1.2   Solar Activity/Flares

The research objective is to understand

- *the solar dynamo.*
- *the precursors to solar activity--the short-term process of active region development and the long-term buildup of polar fields.*
- *the dynamics of magnetic energy buildup in the solar corona and magnetic field topologies, and their role in occurrence of solar flares.*

Solar activity (sunspots, faculae, flares, etc.) arises from the eruption of fields formed within the solar interior as part of dynamo processes. To understand solar activity, it is essential to understand the solar dynamo—how magnetic fields are amplified within the solar interior. To predict future activity, it is essential to understand the precursors of solar activity—on long time scales, the buildup of polar fields, and on short time scales, the behavior of individual active-region development.

Electromagnetic radiation propagates from the Sun to Earth on time scales of minutes. Thus any warning capability for ultraviolet (UV), extreme ultraviolet (EUV), and x-ray or microwave bursts requires some method for flare forecasting. At this time, no completely reliable predictors of flare occurrence are known. Although, because flares are due to the release of magnetic energy, an obvious forecasting strategy is to search for signatures of magnetic energy buildup in the solar corona and specific magnetic field topologies that lead to the occurrence of flares. Recent research has linked sigmoidal signatures to subsequent flare events.

Solar activity can take many forms, ranging in size from tiny x-ray bright points to giant eruptive flares. All these forms of activity are believed to share a common underlying physical process, the conversion of magnetic energy in the solar chromosphere and corona to plasma energy. The best known and, perhaps, most important example of this process is solar flares. A large flare can release up to 10^{32} ergs of magnetic energy into the solar atmosphere on time scales as short as 100 seconds. There are several results of flare energy release that are significant for space weather. The main result is that large masses of coronal and chromospheric plasma heat to temperatures in excess of 10 million degrees Kelvin (K). Radiative and conductive cooling of this plasma produces UV, EUV, and x-ray bursts, from 0.1 to 100 nanometers (nm), which heat and ionize Earth's upper atmosphere and ionosphere.

Flares also produce enhancements of optical emission, especially in spectral lines formed in the chromosphere, but sometimes even in continuum white-light emission. The term flare has now become synonymous with a large soft x-ray burst, and the commonly used flare classification is now in terms of peak soft x-ray intensity. Flares are categorized as X, M, C, or B with the strongest flares being category X. In addition to UV, EUV, and x-ray bursts, flares can produce large numbers of energetic particles that escape to interplanetary space and result in major solar energetic particle (SEP) events. A third important space weather effect of flares is the production of strong radio bursts, although there is considerable ambiguity as to which of these are associated with CME initiation as opposed to the flares themselves.

A.1.3   Solar and Galactic Energetic Particles

The research objective is to understand

- *the origins of high-energy (MeV and GeV) particles and how they propagate through the interplanetary medium.*
- *the related physical processes that modulate the flux at 1 astronomical unit (AU) of cosmic rays originating in galactic space.*

Very energetic particles, whether of solar or galactic origin, can cause single event upsets or latchups in satellite electronic components, especially highly packed memory chips. Although the exact time of failure of individual components cannot be predicted, the overall rate of upsets is proportional to the flux of particles, which is controlled by solar activity and details of the ambient interplanetary medium.

Considering solar energetic particle events, two independent mechanisms for particle acceleration have been identified: processes associated with impulsive solar flares, and shock waves driven by a coronal mass ejection. Particles observed in major solar energetic particle events are accelerated over a period of days over an extended region of solar longitudes, and are associated with interplanetary shocks. The profiles of flux intensity versus time for both these types of events depend on the three-dimensional (3D) topology of the interplanetary magnetic field lines. Research into the particle acceleration mechanisms for both processes is ongoing.

Considering intergalactic cosmic rays, which are more energetic than solar particles, their influx is negatively correlated with the 11-year solar cycle. Apparently, the repeated episodes of solar mass ejections prevalent near sunspot cycle maximum create a complex pattern of interplanetary fields that tend to exclude (scatter) cosmic rays away from the inner solar system. This effect can be seen for individual events; there are obvious decreases in cosmic ray flux at Earth coincident with shock-associated coronal mass ejections.

A.1.4   Solar UV, EUV, and Soft X-Rays

The research objective is to understand the variabilities of the Sun in the short wavelengths and how these variabilities affect the ionosphere and thermosphere.

The upper atmosphere and ionosphere of Earth are subject to extreme spatial and temporal variability induced by solar short-wavelength radiation with wavelengths less than 180 nm, as well as by charged particles precipitating from the magnetosphere and the polar cap, which may be open to interplanetary space. The variability in the solar short-wavelength radiation occurs on time scales from minutes to that of a solar cycle. The ability to predict the often rapid changes in Earth's upper atmosphere and ionosphere rests on our knowledge of the sources of these variations. At present, most of our understanding of the variability of the solar short-wavelength radiative outputs comes primarily from measurements made by the Atmosphere Explorer satellite and several rockets flown in the last two decades.

Variations in solar short-wavelength radiation occur on three basic time scales related to flares, active-region evolution, and the solar cycle. The latter two are emphasized here because flare signature and predictability are essentially covered in section A.1.2. However, it should be repeated that the flare-related enhancements of the short-wavelength radiation can be quite large (over 1000 times background levels) and endure for minutes to hours. Over moderate time scales, say to several solar rotations, short-wavelength solar radiation can show enhancements of as much as 20-30% in response to the emergence, interaction, and evolution of active regions. These variations are further modulated by the passage of the active regions across the visible hemisphere. The variations occurring on these time scales are superimposed on a longer-term modulation of solar radiative output related to and occurring in phase with the solar cycle. During the 3 to 4 years in the rise of an activity cycle, the solar emission in the 10- to 100-nm spectral range can increase by a factor of 2, and for wavelengths less than 10 nm by a factor of 10.

A.1.5   Solar Radio Noise

The research objective is to understand solar radio noise emission processes in the upper corona.

The predominant processes responsible for solar radio noise are related to active regions and their formation and development. Solar radio emission is so closely tied to solar activity that the 10.7-centimeter (cm) flux of solar radio noise is often chosen as a better proxy than sunspot number for solar activity. Nevertheless, proxies are not equivalent to understanding or predicting.

As a factor in space weather, solar radio noise is an operational nuisance and hazard to radar and communication operations if the antenna pattern intercepts the Sun and solar noise swamps the desired signal. The solar radio signals of most interest include noise bursts at 245 megahertz (MHz) and 2.695 gigahertz (GHz) because they interfere with commonly used communication frequencies.

The occurrence rate of solar radio bursts is roughly correlated with the solar activity cycle. However, solar radio noise is also likely simply when there are complex sunspot activity regions in view. Whether this signifies the occurrence of minor flares or slower CMEs with no other identifiable signatures is unknown at this time.

A.1.6   Solar Wind

The research objective is to understand the means by which the solar wind is heated and accelerated, and the non-uniform characteristics of its flow.

Today, advances in understanding the solar wind are being made because of synergistic developments of theory, modeling, and observations. In particular, 3D models, ranging from source-surface treatments to 3D magnetohydrodynamics (MHD) coronal and solar wind models, have the potential to help us understand both how to interpret what we experience at Earth in terms of its coronal origins, and the role that the coronal magnetic field geometry plays in determining the solar wind characteristics. New observations that help to establish the boundary conditions at the Sun have the potential to merge with the models and allow definitive tests of their validity. However, most of these cross-comparisons have not been done. Moreover, the processes acting in the solar wind acceleration region are not completely defined in classical MHD models and are not readily observed by present-day remote sensing techniques.

The solar wind establishes the prevailing condition of the magnetosphere prior to transient (e.g., shock or CME) interaction, is the medium through which solar energetic particles travel (and in many cases are generated in), and is itself a source of moderate interplanetary plasma and field disturbances.

A.1.7   Magnetospheric Particles and Fields

The research objective is to understand

- *the coupling of the solar wind with the magnetosphere.*
- *the onset, expansion, and recovery of substorms.*
- *the transport and energization of plasma throughout the magnetosphere.*

Specifying and forecasting the magnetospheric environment requires a detailed understanding of particles and fields throughout the magnetosphere. Because the magnetosphere is determined largely by properties of the solar wind, a complete understanding of the magnetosphere depends on our knowledge of the physical mechanism or mechanisms behind its coupling to the solar wind. The magnetosphere is a dynamic region that undergoes dramatic changes in response to substorms. Therefore, it is also important to understand the physical processes leading to the onset, expansion, and recovery phases of substorms. Much of the energetic plasma in the magnetosphere resides in the plasma sheet, but our understanding of the origin of the plasma sheet is still incomplete. A significant portion of the plasma sheet ions is provided by the ionosphere, but there is yet no agreement about the mechanisms that transport ions up field lines. These transport mechanisms depend on both storms and substorms in the magnetosphere.

The magnetic field configuration of the outer magnetosphere varies in complex and dramatic ways with changes in the solar wind. On the sunward side of the magnetosphere, Earth's dipole field is compressed by the ram pressure of the supersonically flowing solar wind, and the magnetosphere shrinks when the solar wind ram pressure increases. On the night side, interaction with the solar wind stretches geomagnetic field lines out into an enormous, long tail that extends for hundreds of Earth radii (R_E). A strong southward component of the interplanetary magnetic field (IMF) increases the degree of magnetic connection between the solar wind and magnetosphere, further stretching the magnetic field in the tail. In the expansion phase of a magnetospheric substorm, which typically follows a period of southward IMF, nightside field lines collapse to a less stretched shape, releasing energy. In a magnetic storm not only are there typically strong substorms, but the Earth-centered westward ring current also increases, inflating the magnetic field.

A.1.8 Geomagnetic Disturbances

The research objective is to understand

- *magnetospheric disturbances and the modulating role of the magnetosphere, ionosphere, and neutral upper atmosphere.*
- *the currents that magnetospheric disturbances induce in the ground.*

Magnetic activity at Earth's surface is modified by electric currents in the magnetosphere and ionosphere. Understanding of geomagnetic disturbances depends critically on our knowledge of those currents as well as induced currents in the ground. Areas where greater physical understanding is particularly needed include (1) magnetospheric processes that govern field-aligned currents and conductivity, and produce particle precipitation, especially during substorms and storms; (2) processes that determine rapid spatial and temporal variations of the auroral electrojets (it is the rapid temporal variations that are linked to the ground-induced current effects); and (3) the manner in which spatially and temporally varying ionospheric currents interact with nonuniform ground conductivity to induce large potential differences over long distances.

A.1.9 Magnetospheric Radiation Belts

The research objective is to understand the transport, production, and loss processes that determine the intensity of radiation belt particles in both quiet and storm times.

The radiation belts are the high-energy parts of the populations of electrons and ions that are trapped by Earth's magnetic field. They are distinguished by the energies of their particles, by their spatial properties, and by the variations in time of the fluxes of their particles. They are subdivided into inner and outer belts, separated by a region of minimum particle flux, the slot region, located at about 1.5 to 2 R_E above the equator. The South Atlantic Anomaly is the location where inner belt particles come closest to Earth, owing to the offset between the magnetic dipole axis and Earth's center. The outer belt extends beyond the altitude of geosynchronous satellites.

Although typically thought of as regions of trapped radiation, the radiation belts exhibit variations on nearly all time scales: secular, solar cycle, solar rotation, and storm time. The outer belts exhibit the greatest variability. The outer belt MeV electrons, often called "killer electrons" because of their effect on spacecraft, exhibit a persistent behavior during magnetic storms, yet their origin is still not understood. Even the inner belt particle populations, which are usually considered stable, can be modified during large magnetic storms. For example, in 1991 an interplanetary shock wave created a second inner proton belt, which, because of the long time scales associated with loss processes, persisted for many months.

The transport, production, and loss processes that determine the fluxes of energetic particles in the radiation belts have been studied for many years, but it is still not possible to account for all the observed variability. Transport, production and loss of radiation belt particles result from atmospheric processes such as charge exchange and Coulomb collisions, pitch angle diffusion produced by wave-particle interactions, and radial diffusion caused by large-scale magnetic and electrostatic impulses. The determination of numerical values for the various diffusion coefficients is a critical step for understanding and predicting the evolution of the radiation belts. This will require a detailed specification of the electromagnetic and electrostatic fields present in the radiation belts, as well as a suitably robust model of the large-scale magnetic field.

A.1.10   Aurora

The research objective is to understand the processes that guide, accelerate, and otherwise control particle precipitation and produce auroral substorms.

Much has been learned over the past two decades about the climatology of auroral particle precipitation. As we understand more about where magnetic fields map, we are better able to tie features in the precipitation patterns to source regions in the magnetosphere and magnetosheath and to processes that guide, accelerate, and otherwise control particle precipitation. Adding the dependence of the climatology on the IMF direction may help to define the physics of these processes. Critical questions still exist relative to the acceleration mechanisms for auroral particles and the causes of structure of arcs and other auroral forms. We must better understand magnetosphere-ionosphere coupling and feedback processes. The dynamics of the aurora reflect the extreme temporal variability of the magnetosphere and the coupling processes. Many of the research models are run under steady boundary conditions to produce diurnally reproducible results. Time-varying effects are often not well modeled. Yet the solar wind, the magnetosphere, and the ionosphere vary considerably on time scales of hours and even minutes. On the global scale we must evolve from static, statistically based pattern descriptions of electric fields, currents, and precipitating particles to dynamic time-varying real-time descriptions of the physical processes controlling these phenomena. Accurate simulation of time-varying phenomena at scales of a hundred kilometers to across the globe is an essential part of space weather prediction.

A related problem is the triggering mechanism of substorms and the modeling of the dynamics of the expansion phase, including the processes that lead to the "injection" of intense fluxes of energetic particles into the near-geosynchronous region. There is a lack of agreement in the community on the physical processes that cause substorm initiation, although focus has moved toward the inner magnetosphere in the last few years. It is ultimately important that the physics of substorm onset be identified and be the driver of onset in numerical models rather than any ad hoc mechanism or numerical diffusion within the model.

A.1.11   Ionospheric Properties

The research objective is to understand

- *the formation mechanisms associated with large-scale and medium-scale electron density structures, and the basic response of the ionosphere to geomagnetic storms and substorms.*
- *the production, transport, and loss mechanisms associated with electron density structures.*

Earth's ionosphere displays a marked variation with altitude, latitude, longitude, universal time, season, solar cycle, and magnetic activity. This variation is reflected in all ionospheric properties: electron density, ion and electron temperatures, and ionospheric composition and dynamics. This is primarily a result of the ionosphere's coupling to the other regions in the solar-terrestrial system, including the Sun, the interplanetary medium, the magnetosphere, the thermosphere, and the mesosphere. The main source of plasma and energy for the ionosphere is solar EUV and UV radiation, but magnetospheric electric fields and particle precipitation also have a significant effect. The magnetospheric effect is determined, in part, by the solar wind dynamic pressure and the orientation of the interplanetary magnetic field (IMF). Also, tides and gravity waves propagating up from the mesosphere influence the thermospheric neutral densities, which, in turn, affect electron and ion production and loss rates. The various driving mechanisms act in concert to determine the global electron density distribution, but important time delays and feedback mechanisms are also associated with the coupling processes. The external driving mechanisms can also be localized, spatially structured, and unsteady.

Of particular importance to space weather systems is the electron density distribution. Despite the complicated nature of the forcing processes, this distribution exhibits (generally) repeatable features at equatorial, middle, and high latitudes. At mid-latitudes, the average electron density distribution tends to be uniform, with a gradual transition from dayside high densities to nightside low densities across the terminator. At equatorial latitudes, a pronounced latitude variation of electron density in the F-region of the ionosphere, known as the Appleton anomaly, is encountered during the daytime. At high latitudes, additional large-scale density features are evident, including a tongue of ionization, a polar hole, a main trough, and an overall enhancement in the auroral oval. In addition to these quasi-steady-state features, the ionosphere also exhibits a considerable amount of structure. There are small-scale (~1 km), medium-scale (~10 km), and large-scale (100-1000 km) density structures, and they can appear at any location and time. The small-scale structures are usually produced within and on the edges of the larger structures through plasma instabilities and are typically referred to as density irregularities. The medium- and large-scale structures at high latitudes can appear in the form of propagating plasma patches, boundary blobs, auroral blobs, and localized depletions, and they can be created by a variety of mechanisms. At the magnetic equator, large-scale features known as equatorial bubbles or plasma depletions appear after sunset and are associated with large-amplitude density irregularities at a variety of scale sizes.

A.1.12   Ionospheric Electric Fields

The research objective is to understand

- *the small-scale electric field (E-field) structures and the large-scale electrostatic fields within the ionosphere, how they couple with the magnetosphere, and how they respond to changes in the interplanetary magnetic field.*
- *the penetration of E-fields from high latitudes to low latitudes.*
- *the E-field variability generated by thermosphere-ionosphere interactions in the equatorial region.*

A specification of the global E-field is required to adequately understand the behavior of the charged and neutral species in the upper atmosphere. The E-field forces the ions to move in the $\mathbf{E} \times \mathbf{B}$ direction at high altitudes and in directions closer to the E-field direction at lower altitudes. The resulting plasma transport and ion-neutral chemistry affect the ionospheric composition and concentration, which, in turn, affect the neutral atmosphere dynamics by modifying the ion drag. These processes are further complicated by the fact that the ion-neutral collisions themselves drive currents that may modify the E-field and the resulting $\mathbf{E} \times \mathbf{B}$ drift motion of the plasma. To produce an adequate description of the ionosphere and thermosphere, it is therefore necessary to globally specify the E-field distribution. At the large spatial scales that are relevant to describing the global ionosphere, the magnetic field lines are electric equipotentials, and thus the E-field may be specified by describing the spatial distribution of electric potential. Small-scale structure that is superimposed on the large-scale E-field is also important. The task of describing the global ionospheric E-field may be conveniently divided into two parts: that pertaining to the specification at high latitudes, where the influences of ionosphere-magnetosphere coupling processes dominate, and that pertaining to middle and low latitudes, where ionosphere-thermosphere-mesosphere coupling processes dominate.

At high latitudes the ionospheric E-field is largely dependent on the nature of the magnetosphere-solar wind interactions, and interplanetary magnetic field and solar wind parameters tend to be the major variables affecting the electrostatic potential distribution. When the IMF is southward, a two-cell convection pattern is generally well defined, but the level of detail required to adequately specify the associated ionospheric densities is missing. When the IMF is northward, the variability of the convection pattern is much larger and the convection velocities are generally smaller than for the southward IMF case. This suggests that the underlying large-scale convection pattern may be disguised by the small-scale structure that is present in all cases. At low and middle latitudes, a specification of the E-fields is complicated by the existence of daily and longitudinal variations that are poorly understood. The relationship between the behavior of the ionosphere and thermosphere and the electrodynamics of the plasma can be quite well modeled, but prediction of the wind systems and E-field variations that can dramatically affect the bottomside ionosphere is not presently possible. A better understanding is required of the processes that determine the spatial and temporal characteristics of E-field penetration from high to low latitudes during ionospheric disturbances, including the establishment or breakdown of magnetospheric shielding effects.

A.1.13   Ionospheric Disturbances

The research objective is to understand the day-to-day variability of the large-scale ionospheric features and small-scale plasma density irregularities that affect radio wave propagation during magnetically quiet and disturbed times.

A multitude of structures populates various ionospheric regions at different times. Their spatial and temporal distributions are of interest in their own scientific context but are also of interest to communication and surveillance applications because of their impact on skywave signal channel characteristics. At mesoscale and macroscale (i.e., ~ 50-10^4 km) the interest has been in such phenomena as the auroral oval, the equatorial anomaly, the mid-latitude trough, the cusp, intermediate and descending layers, and polar cap patches. At smaller-scale sizes (centimeters to tens of kilometers) attention has been on plasma instabilities that play a role in the distribution of ionospheric irregularities. Through wave-particle interactions the instabilities and associated irregularities can influence currents and energize ions that ultimately populate the magnetosphere. Many processes (Rayleigh-Taylor, $\mathbf{E} \times \mathbf{B}$, current convective, and universal drift wave modes) depend on ionospheric structures and their density gradients as energy sources to drive the plasma to an unstable state. All of the high-latitude ionosphere and the nighttime equatorial region are susceptible to such processes. These regions are known to impair skywave performance when such instabilities and irregularity structures prevail. It is to be noted, however, that ionospheric disturbances are not always associated with instability mechanisms. They can be triggered by geomagnetic storms and attendant variations in E-fields, thermospheric winds, composition, and plasma density distributions.

Triggering mechanisms can also include gravity wave perturbations and day-to-day variabilities that can often escape identification with known cause-effect relationships. In fact, an underlying challenge (perhaps more fundamental than attempting to trace storm-time dynamics) is to define and specify quiet-time conditions (if such a classification truly exists) and their associated day-to-day variability. Herein lies the fundamental requirement for skywave performance characteristics in ionospheric controls of maximum usable frequencies, lowest usable frequencies, and frequencies of optimum transmission. These are link dependent and controlled by the details of E- and F-region characteristics as determined by seasonal, diurnal, and solar-cycle controls, and attendant day-to-day variability. Clearly, an understanding of all structures, laminar and disturbed, will provide a hierarchical perspective on all of the ionosphere and will develop insights into the processes that control and maintain the structures themselves, influence the coupling to other geospace domains, and modify the performance of communications and surveillance systems.

A.1.14   Ionospheric Scintillations

The research objective is to understand

- *the thermosphere-ionosphere-magnetosphere interactions that control the formation and evolution of 10-km to 50-m electron density irregularities that cause scintillations.*
- *the relationship between those irregularities and scintillation effects on specific systems.*

Spatial irregularities of ionospheric electron densities scatter satellite radio signals and lead to amplitude and phase variations. Amplitude scintillations induce signal fading and, when this exceeds the fade margin of a receiving system, message errors in satellite communications are encountered and loss of lock occurs in navigational systems. Phase scintillations cause Doppler shifts and may degrade the performance of phase-lock loops, such as in Global Positioning System (GPS) navigation systems. They may also affect the resolution of space-based synthetic aperture radars. The magnitudes of amplitude and phase scintillations and the temporal structure of scintillations need to be specified and predicted to provide support to operational communication and navigation systems.

The scintillation phenomenon is characterized by extreme temporal and spatial variability. The associated electron density irregularities are driven by complex, time-dependent ionospheric plasma instabilities. Research into the triggering mechanisms for the instabilities is crucial for achieving predictive capabilities. Scintillations are most severe in the equatorial region, where they often occur after sunset, and attain their maximum intensity around the peaks of the Appleton anomaly (15 degrees north and 15 degrees south latitude, magnetic). The scintillations have a large and poorly understood day-to-day variability, with active nights sometimes following (or leading) quiet nights with little apparent change in initial conditions. They are particularly severe during solar maximum conditions, and frequently occur during geomagnetically quiet periods.

At high latitudes, strong scintillation events are related to the macroscale plasma structures that become unstable near their moving boundaries. Under magnetically active conditions and IMF B_z southward orientations, such structures, known as polar cap patches, are convected from mid-latitudes through the dayside cusp into the polar cap and finally into the nightside auroral oval. It is first necessary to specify and predict polar cap patches and their trajectories based on IMF configurations. We then need to define plasma convection in the neutral frame of reference so that the growth time of plasma instabilities and relative amplitude of mesoscale irregularities may be derived in the nonlinear regime of the evolution of plasma instabilities. At mid-latitudes, weak to moderate levels of scintillation occur and maximize during the solar minimum period in a manner that still awaits explanation.

The physical conditions necessary for the onset of plasma instability in the equatorial region seem to be associated with the post-sunset enhancement of the eastward E-field and the presence of "seed" perturbations, provided by either geophysical noise or gravity wave activity. However, the manner in which these parameters control the day-to-day variability of scintillation remains unresolved. For the purpose of relating scintillations to plasma instabilities on a quantitative basis, it will be necessary to track these instabilities through the nonlinear regime and determine the saturation amplitudes of the irregularities with fast computer algorithms.

A.1.15   Neutral Atmosphere (Thermosphere and Mesosphere)

- *The research objective is to understand the chemical, radiative, and dynamical processes that act to modify and redistribute energy and constituents throughout the upper atmosphere.*

The atmosphere above about 80 km is a weakly ionized, compressional, multi-constituent medium with a complex and variable morphology that is controlled by a variety of mechanisms, including direct solar heating and other important chemical, radiative, and dynamical coupling processes. Solar activity influences upper atmospheric variability directly by means of EUV/UV heating and indirectly by means of the magnetospheric sources of energy, momentum, and mass. Consequent variations in the neutral atmospheric state parameters (temperature, density, wind, and composition) have important effects on many operational systems. Direct effects include the perturbation of satellite trajectories in low Earth orbit. Indirect effects involve the response of the ionosphere to variations in neutral atmospheric composition and dynamics. For example, changes in thermospheric composition control ionization and recombination rates, whereas neutral winds act to redistribute the ionospheric layers both horizontally and vertically. Neutral winds can produce feedback on the magnetosphere-ionosphere electrical circuit by modifying the dynamo interaction. Also, neutral temperatures play a role in determining chemical reaction rates, which, in turn, affect all other parameters in the upper atmosphere.

Although sustained progress in understanding the upper thermosphere has occurred over the past decade and most of the important physical processes are known, we still possess only a preliminary climatology. Moreover, large errors regularly occur when too much reliance is placed on the accuracy of empirical models. This is due, in part, to the large intrinsic variability of the neutral upper thermosphere and its ability to support oscillations of all types and scales (gravity waves, planetary waves, and tides). Much less is known about the behavior of the lower altitude regions: the mesosphere and lower thermosphere/ionosphere (MLTI).

The MLTI is subject to strong forcings from both above and below, in the form of solar radiation; precipitating magnetospheric particles; magnetospheric electric fields; and currents, atmospheric waves of all scales, radiative transfer, and the transfer of important neutral constituents from the lower atmosphere. Many of the most important atmospheric processes in this region remain poorly characterized, including: the effects of breaking gravity waves; the variability of tides and planetary waves; the production, transport, and loss of species such as nitric oxide and atomic oxygen; radiation cooling under conditions of strong non-thermodynamic equilibrium; and heating by solar radiation, particle precipitation, chemical transformations, and auroral electric currents.

A.2   Model Development

The NSWP must support the continued development of models that specify and forecast the state of the solar-terrestrial system. The program must also foster the merging and integration of models developed for the different space weather regions. As the models are developed, they must continually be tested and validated against space weather measurements. This will necessitate both rapid access to existing ground- and space-based observing platforms as well as deployment of new facilities and satellites.

Different models are in different stages of maturity. Some are empirical, some are "physics based"; most are hybrids. The approach of the NSWP is to support the evolution from empirical models to coupled physical models. In the ideal situation, all models are self-consistently coupled so that changes in the driving forces of the system are communicated through the different regions of space. Finally, the physical models must not only be brought into existence but they must also be transferred into operational codes that will allow accurate space weather forecasts and nowcasts.

A.2.1   CME Models

CMEs produce dramatic effects in the solar wind that must be accurately modeled to predict the onset times and time profiles of solar wind disturbances at 1 AU (plasma, magnetic field, and energetic particle attributes, including strength of associated shocks, intensities of energetic particle events, flow speeds, densities, and magnetic field magnitudes and orientations). Such predictions should eventually be possible on a variety of time scales ranging from less than an hour to greater than several days. The optimum means of achieving this goal is through a combination of observations and models. The observations are required both to test the models and to provide the boundary conditions for model use in making forecasts. Of particular value in the modeling area for CME space weather effect predictions are the following:

- 3D MHD models of the ambient solar wind (see section A.2.4).
- 3D MHD models of CME-generated disturbance propagation in the solar wind from the Sun to beyond Earth's orbit. These models should strive to simulate the CME structure itself as well as the perturbation caused by the CME in the ambient interplanetary medium. They should ultimately be able to describe disturbance initiation and propagation from the base of the corona to 1 AU using realistic initial conditions for the ambient wind and realistic boundary conditions for the CME disturbance itself.
- Models of the CME-driven shock-related radio emission process. These models are needed to optimize the use of radio noise as a remote sensing device and as a diagnostic of approaching CMEs.

A-14

- Models of the CME initiation process that use realistic observable boundary conditions at the Sun to predict "injection" speed, mass, and intrinsic magnetic field attributes of the ejecta.
- 3D models of particle acceleration by CME-driven interplanetary shocks. Given a realistic model of the disturbance propagation (as in the second bulleted item above), these models should be capable of predicting the intensity and time history of the CME-associated energetic particle events at 1 AU.

A.2.2   Flare Models

The magnetic field plays a fundamental role in the production of flares and is the essence of several areas of flare modeling. First, because the coronal magnetic field cannot be observed directly, as yet, the only method for determining the field there is to observe it at the photosphere and use numerical modeling to extrapolate it into the corona. This method has had considerable success in matching observations from present magnetographs and the Yohkoh satellite. Because much higher quality data is expected in the next few years, magnetic extrapolation models must be extended to much higher numerical resolution. Second, the fundamental process by which magnetic free energy in the corona is released is believed to be magnetic reconnection. Recently, 2.5D models for magnetic reconnection have been successful in explaining observations of mass acceleration and heating associated with the magnetic energy release in chromospheric explosions. Solar flare reconnection, however, is intrinsically 3D. Rigorous models for 3D reconnection are the most important challenge to understanding flare physics, and to obtaining a physics-based model for flare prediction. Related to this, investigation of the microphysics that determines the site of reconnection in a particular magnetic field configuration is useful in assessing the potential for reconnection. These processes can generally be parameterized in MHD models to control the locations of the regions of maximum resistivity, and thus minimize the role of purely numerical diffusion.

Models relating to the processes that produce flare-generated transient UV, EUV, and x-ray bursts are potentially valuable for flare effect prediction because they can tell us something about the energetics of the flare and in particular about the potential for producing flare-generated solar energetic particles. Similarly, models of the processes that accelerate particles in flares and determine whether and where they are released from the flare site are useful in building forecasting schemes for this particular component of the SEP population.

The key modeling objectives for flare forecasting are thus the following:

- Development of 3D models for magnetic reconnection in active regions, including consideration of the processes that determine the distribution and magnitude of resistivity.
- Improvement of magnetic field extrapolation models based on photospheric field measurements, allowing for anticipated high-resolution magnetograph observations and eventually higher sensitivity full-vector magnetic field data.

A-15

A.2.3   Solar UV, EUV, and Soft X-Ray Models

Models of Earth's ionosphere and upper atmosphere rely on the solar 10.7-cm radio flux as a surrogate or proxy of the solar short-wavelength radiative input. The variation in the 10.7-cm radio flux, however, originates in active regions. Recent studies show that a significant contribution to the short-wavelength radiation emission levels comes from the dispersed fields of active regions and the weak network. This component of solar activity is not well represented by the 10.7 cm flux. The Ca II K index and He I 1083-nm equivalent width include contributions from active regions, as well as the weaker magnetic structures distributed over the entire surface of the Sun. However, better estimates of the short-wavelength radiative emission can be realized by using solar atmospheric models for different activity structures, such as sunspots, plages, active regions, and the network. In addition, understanding the physical basis for the variability and improving the development of reliable proxies are valuable.

The most promising research directed at this problem involves combining semi-empirical atmospheric models that fit the spectra of observed spatially resolved solar features, databases of line and continuum spectra computed in local thermodynamic equilibrium (LTE) and non-LTE, as appropriate, and analyses of the distribution on the solar disk of specific activity features from observations. This procedure yields a synthetic disk image, full-disk spectra, and the absolute full-disk irradiance in selected spectral bands. At the present time, the solar models are limited to lines formed at temperatures less than 10^5 K, that is, to the chromosphere up to the lower boundary of the transition region. Comparisons of the calculated Lyman-alpha irradiance, for example, with that measured by the Upper Atmosphere Research Satellite (UARS) show an agreement to within about 10%, indicating the potential of this analysis approach.

Specific modeling tasks include the following:

- Develop 3D models of the solar atmosphere that include the transition region and coronal lines formed at higher temperatures than are currently being used.
- Improve the solar spectrum calculations with the addition of line opacities in approximately 50 million lines, mostly in the UV, as well as other lines and continua from atoms, ions, and molecules. Improve non-LTE computations of metals continua, Fraunhofer lines, and the optically thin transition region and coronal lines.
- Develop additional models (e.g., for the penumbra), a finer distinction of modeled structures, and dynamical models that include flows and turbulent diffusion in order to better match the observed solar features and their spectral characteristics.

A.2.4   Solar Wind Models

Over the past two decades numerical models of the solar wind of various degrees of sophistication have been developed. The simplest of these are kinematic models that assume the magnetic field and velocity at the Sun and project them to 1 AU. Some of these models incorporate ad hoc descriptions of stream interaction effects. At a more computationally demanding and physically rigorous level are 2D and 3D MHD models that can in principle more accurately simulate stream structure. Solar observations, especially of the photospheric magnetic fields, can be used to specify the boundary conditions at the "source surface," although theoretical extrapolations or approximations are required to map these conditions to the ~20 solar radii (R_S) distance where the flow becomes truly radial and the speed is thought to be fully established. From a solar wind forecasting perspective, modeling efforts will eventually lead to a capability for predicting local solar wind conditions that can be used in magnetospheric and upper atmospheric models that depend on solar wind behavior, as well as in predictions of CME effects on the ambient interplanetary medium. Because these models rely on solar observations for their boundary conditions, the lead-time is potentially no shorter than the convection time from the Sun and is as long as a solar rotation if the solar magnetic field configuration is steady or slowly changing. Some examples of specific modeling activities of interest in this area are as follows:

- Development of 3D MHD models of the coronal acceleration region of the solar wind, which use realistic magnetic field configurations.
- Development of 3D MHD models of the solar wind extension into interplanetary space, which use realistic inner boundary conditions. These include using the critical "geoeffectiveness" parameters of plasma velocity and density, and vector magnetic field.
- Coupling of the above two models to determine an optimum proxy for specifying solar wind velocity prior to its free expansion into interplanetary space.

A key element in forecasting geomagnetic disturbances is the availability of solar wind and IMF data from the Lagrangian point (L1). Since the L1 point lies more than 200 R_E upwind from Earth, and since an L1 station is rarely on a streamline that hits the magnetosphere, there is a need to predict solar wind and IMF conditions at the magnetosphere from the L1 data. Existing models make such predictions using pure advection based on the assumption that all gradients are parallel to the Sun-Earth line. It should be possible to improve the accuracy of the predictions by computing the orientation of gradients from the L1 data and taking account of this information together with the propagation of the gradients relative to the wind. Two real-time L1 data stations are WIND and the Advanced Composition Explorer (ACE). Codes incorporating data from both stations should give accurate predictions most of the time. The extension of existing code to incorporate these aspects of the solar wind transit process should be completed and tested.

A-17

A.2.5   Magnetospheric Particle and Field Models

Magnetospheric models are needed to specify and predict the particles and fields in the magnetosphere and radiation belts, as well as the electrical currents that produce geomagnetic disturbances at Earth's surface. All magnetospheric models depend on knowledge of the solar wind, and the lead-time for magnetospheric predictions can only be as good as the corresponding lead-time for the solar wind inputs.

Currently, a magnetospheric model called the Magnetospheric Specification Model (MSM) is being used operationally to estimate fluxes of electrons and H$^+$ and O$^+$ ions in the distance range from L ~ 2 to nearly the magnetopause on the dayside and to about 15 to 20 R$_E$ on the nightside. The MSM has at its core a particle drift code, with electric and magnetic field models that are driven by real-time data. The MSM can specify magnetospheric conditions on the basis of the Kp three-hourly planetary index of geomagnetic activity alone, but the quality of its output increases as its input database increases. The MSM is limited to providing nowcasts and retrospective analyses. However, the slightly more advanced Magnetospheric Specification and Forecast Model (MSFM), which is now in limited operational use, provides forecasts with a typical horizon of about 1 hour, as set by the availability of solar wind and IMF data from an L1 station.

The MSFM represents only one approach to numerical magnetospheric prediction codes. A principal shortcoming is the lack of quantitative, detailed self-consistency between the particle populations it numerically calculates and the magnetospheric magnetic field that it gets from a lookup table. Such self-consistency between plasma and magnetic field is the strength of global MHD simulations, which represent an alternative approach to numerical magnetospheric predictions. A principal shortcoming of global MHD simulations is their neglect of thermal drifts, which are important where field gradients are strong. Thermal drifts are explicitly calculated in the MSFM. Clearly a merger of an MSFM-like code (i.e., containing non-MHD physics) and a global magnetospheric MHD code represents a goal that should be set in order to advance toward greater capability in numerical space weather predictions. NASA has a plan, called the Quantitative Magnetospheric Predictions Program, to develop an operational numerical magnetospheric prediction capability based on integrating non-MHD physics into a global magnetospheric MHD code. At the same time the Department of Defense (DOD) is supporting the development of a particular version of a merger between an MSFM code and a global magnetospheric MHD code.

The present versions of the MSM and MSFM require accurate specifications of the electric and magnetic fields and the ionospheric conductance produced by auroral precipitation. One approach is to couple the MSFM to models of the ionospheric electrodynamics that assimilate data from ground-based magnetometers and radars, as well as from polar-orbiting satellites. Alternatively, semi-empirical models for E-fields and magnetic fields and auroral particle precipitation can be used. Models for E-fields and auroral precipitation are described in sections A.2.10 and A.2.8, respectively.

Various semi-empirical and theoretical models of the magnetospheric magnetic field are currently being developed. Semi-empirical magnetic field models are partly based on observations, but they also generally conserve magnetic flux and satisfy Ampere's Law. However, they do not explicitly relate currents and magnetic fields to plasma pressures, densities, or velocities. These models are usually designed to be user-friendly, and they run quickly on workstations. Theoretical models, which include the additional theoretical constraint of momentum conservation (and often conservation of mass and energy as well), can represent dynamical processes better, but they can run faster than real time only on the most advanced supercomputers or massively parallel machines, and they are not yet ready for operational use.

When the MSFM is successfully merged with a global magnetospheric MHD code, auroral precipitation E-fields and magnetic fields will be self-consistently computed. However, the semi-empirical models should continue to be refined and developed to validate and support the fully-coupled model.

A.2.6   Geomagnetic Disturbance Models

For some technical systems, it is not necessary to know the particles and fields everywhere in the magnetosphere. In some cases, it is sufficient to specify and predict the level of geomagnetic disturbance on the ground, or what is referred to as geomagnetically induced currents (GIC). A number of statistical algorithms are being developed to specify and forecast geomagnetic disturbances. These include an input-state space algorithm for the auroral electrojet (AE) index and more than one neural network algorithm for the magnetic disturbance storm time index (Dst). A statistical auroral electrojet predictor that specifies when, where, and how much the electrojet will intensify is being developed for use in making GIC warnings. These and similar algorithms that take the powerful statistical techniques that have been developed in the fields of nonlinear dynamics and artificial intelligence and use them to predict geomagnetic disturbance indices constitute a potentially rich source of useful specification and forecast products that can be transferred into operational service in the near term. Although prediction in the long term will be best done by physics-based numerical modeling, new techniques are under development to apply locally linear prediction filtering (of nonlinear phenomena) and deterministic chaos theory to characterize the magnetospheric response. Development of these and similar techniques could provide an alternative means for forecasting substorm activity through prediction of indices such as AE until the physics of substorm onset is better understood.

One deficiency all statistically based algorithms share is their poorly understood performance characteristics during extreme conditions, for which there are few data with which to fit or train them. Nonetheless, the mere fact that an algorithm is predicting out of range can have considerable forecast value. Regarding observational requirements, as a rule, these algorithms need real-time values of the index or quantity they predict, and some also need real-time upstream solar wind data.

A.2.7   Radiation Belt Models

Radiation belt particle populations have historically been specified by National Aeronautics and Space Administration (NASA) empirical models for inner and outer zone protons, called AP-8 Min and Max, and for inner and outer zone electrons, called AE-8 Min and Max, where Min and Max refer to phases of the solar cycle for which the models were designed. These models are based on data acquired before about 1990. All static models show various inaccuracies, especially in the South Atlantic anomalies and in the outer zone. Even in the more stable inner zone there is a deficiency in specifying >100 MeV protons and >10 MeV electrons.

Data from the Combined Release and Radiation Effects Satellite (CRRES) and Solar Anomalous Magnetospheric Particle Experiment (SAMPEX) satellites have demonstrated the extremely dynamic nature of the radiation belts and the inadequacy of the NASA AP-8 and AE-8 models. Outer zone relativistic electrons exhibit strong variations that are well correlated with the solar cycle, solar rotation, and solar wind speed. Inner zone protons can suffer sudden changes during magnetic storms and such changes can persist for months. This deficiency of the present models calls for research into dynamical models. Examples of such models have already appeared. The sudden creation by an interplanetary shock wave of a second inner belt has been successfully modeled. A neural network model based on Kp gives reasonably good predictions of the presence of relativistic electrons at geosynchronous orbit. Newer models of particle fluxes based on data from the CRRES and SAMPEX satellites are now becoming available. New CRRES radiation belt models which specify proton flux, electron flux, and total radiation dose are replacing the NASA models as inputs for specifying the radiation environment for satellite design engineers. The potential for improving and extending these and other models for operational purposes is great.

A.2.8   Auroral Models

An important element for both ionospheric and magnetospheric models is the ability to specify and predict the location and intensity of auroral precipitation. Current modeling efforts combine statistically based empirical models with known physics to produce reasonable assessments of inner magnetosphere dynamics and the resulting particle precipitation into the ionosphere. The MSFM provides the first numerical capability to follow magnetospheric particle fluxes and determine the precipitation into the auroral zone and an estimate of the time-varying electric field. Predictions of up to 1 hour are possible with solar wind data for input. However these models are missing good descriptions of time-varying, magnetosphere-ionosphere coupling and the resulting particle acceleration and field-aligned currents. The effects of the magnetospheric energy on the ionosphere can also be estimated with regionally based models that have no magnetosphere-ionosphere coupling.

A.2.9   Ionosphere Models

Because of the complicated nature of the ionosphere, there have been numerous approaches to ionospheric modeling over the years. These approaches include the following: (1) empirical models based on extensive worldwide data sets; (2) simple analytical models for a restricted number of ionospheric parameters; (3) 3D, time-dependent physical models including self-consistent coupling to other solar-terrestrial regions; (4) models based on orthogonal function fits to the output obtained from numerical models; and (5) models driven by real-time magnetospheric inputs. In an effort to achieve simplicity, some of the models have been restricted to certain altitude or latitude domains, while others have been restricted to certain ionospheric parameters, such as NmF2 and hmF2. Most of the models have been constructed to describe the climatology of the ionosphere and, in this regard, the models, such as the International Reference Ionosphere (IRI), have been very successful in describing the characteristic ionospheric features and their variations with universal time, season, solar cycle, and geomagnetic activity, as represented by Kp and Ap. More recently, the model development has focused on including the large-scale and medium-scale density structures in global simulations in a self-consistent manner. Efforts have also been directed toward modeling storms and substorms.

Global approaches to ionospheric weather and climatology include the Thermosphere-Ionosphere Electrodynamics General Circulation Model (TIEGCM), the Time-Dependent Ionospheric Model, and the Field-Line Integrated Plasma model. Currently, the Parameterized Real-time Ionospheric Specification Model (PRISM) is being used to provide a near-real-time specification of the global ionosphere by starting with a parameterized physical model of the ionosphere and adjusting it to match near-real-time ionospheric data. The output of theoretically pre-calculated electron density profiles was parameterized to achieve sufficient computational speed. The near-real-time data ingested by PRISM are obtained from a network of ground-based sensors (bottomside soundings using digital ionosondes and GPS-based total electron content measurements) and space-based measurements from the suite of sensors on two Defense Meteorological Satellite Program (DMSP) satellites (in situ plasma data and precipitating ion and electron fluxes). PRISM is operational at the Air Force Weather Agency's 55th Space Weather Squadron (55 SWXS), providing global electron density profiles from 90 to 1600 km every 2 degrees latitude and 5 degrees longitude.

Critical parameters must be available either by means of measurements or models to improve ionospheric nowcasts. At high latitudes, the convection E-field and particle precipitation patterns must be known. Unfortunately, these 2-D patterns are needed as a function of time for dynamic simulations. It is, therefore, not a surprise that the bulk of the ionospheric modeling conducted to date pertains to climatology, because "empirical" or "statistical" E-field and precipitation models are appropriate in this case. Here, the empirical convection and precipitation patterns are held fixed for a 24-hour period, and diurnally reproducible electron densities are calculated. In the first few attempts to model time-dependent phenomena, such as geomagnetic storms, empirical convection and precipitation patterns were also adopted, but they were varied in time according to the variation of Kp with time. More recently, the time-dependent convection and precipitation patterns have been obtained by an ionospheric electrodynamic model driven by magnetometer, radar, and satellite data. With regard to the other ionospheric regions, the meridional wind is the critical parameter at mid-latitudes and the dynamo electric field is crucial at low latitudes. These parameters can be deduced by means of measurements of hmF2 or they can be calculated self-consistently using coupled ionosphere-thermosphere-electrodynamic models.

Among the first-principle models, deficiencies exist in both the topside and bottomside regions of the ionosphere. On the topside, the major problem exists at night and involves the plasmaspheric H^+ flux and its control of F-region heights and densities. Inability to accurately specify this flux is known to result in major errors ($\sim>100\%$) in predictions of nighttime F-region densities. Until this problem is resolved, the relative roles of thermospheric winds and electric fields in the maintenance of the nighttime F-region will remain an open issue.

On the bottomside, it is necessary to specify the global distribution of upward-propagating tides and gravity waves, which is generally unknown. Major deficiencies in first-principle models also include inabilities to accurately specify the E-region and F1-region, including the distributions of intermediate, descending, and sporadic-E layers. These layers have gained attention because of their relevance to dynamo fields, their kinetic interactions with thermospheric winds, their enhanced conductivity and associated controls of E-region current systems, and their potential role as a tracer of wind-shear nodes and tidal components. Model deficiency in specifying these layers stems from inabilities to accurately specify NO, NO^+, and metallic ion populations along with forces due to zonal and meridional winds.

It is important to develop nested-grid, adaptive-grid, and nested-model approaches so that density structures of various scales can be self-consistently included in global simulations. In the short term, continued improvement in empirical electron density models is required, based on the new databases. For the long term, it is important to further develop coupled physical models. In addition, computationally fast empirical-numerical hybrid models will eventually be needed in order to develop a real-time forecasting capability. As part of this effort, multi-site data taken in real time will be needed for ingestion into the forecast models.

A.2.10   Ionospheric Electric Field Models

At high latitudes the ability to specify the global potential distribution lies in four areas: analytical or semi-analytical models driven by interplanetary parameters; adaptive numerical models driven by observations; data assimilation techniques driven by observations; and global magnetospheric models driven by interplanetary parameters. Evidently a predictive capability for the high-latitude potential distribution lies in improvement of the analytical or semi-analytical models and the refinement of global magnetospheric models. Immediate improvement of the present E-field nowcasting capability may be accomplished by improvement in data assimilation techniques and in the refinement of the adaptive models. The results from data assimilation procedures need to be quantified in a manner that makes comparison and integration into analytical models more straightforward.

At low and middle latitudes, a specification of the E-field is largely available from statistical analysis of data sets from incoherent-scatter radar sites rather sparsely distributed around the globe. Incorporation of these latest data sets into a global E-field model has not yet been undertaken. Work continues on the effects of high-latitude phenomena on the low-latitude ionosphere, with quantitative estimates of the penetration and disturbance effects to be expected in the near future to allow transition to a forecast model. Accurate specification of the longitude variations requires a wider distribution of measurements that are not presently available. The height of the F-region peak density and the distribution of ion concentration about the dip equator, are sensitive functions of the $\mathbf{E} \times \mathbf{B}$ drift motion of the plasma. It is thus possible that a description of these plasma properties may serve as a proxy for the $\mathbf{E} \times \mathbf{B}$ drift motion. In general, advances in this area are currently hampered by the lack of contiguous data sets that would allow accurate assessment of daily variations and their possible sources.

The goal of achieving some convergence in the present specification of the high-latitude potential distribution from each technique should be achieved in the near future. Results from global numerical models can now be compared with observations, and we may expect refinements in the models to occur in response to such comparisons. As a far-future goal, we should expect a convergence in the specification of E-fields from all these techniques and an increase in the efficiency of global numerical models, allowing their more frequent use. Models now generally provide a specification of the high-latitude potential distribution for a rather coarse profile of interplanetary parameters. The challenge is to provide a more continuous nowcast and forecast capability and the appropriate methodology to evolve the pattern from one state to the next.

A.2.11   Ionospheric Disturbance Models

Currently, ionospheric models do not incorporate ionospheric disturbances that affect the propagation and transmission of radio waves. These disturbances are associated with phenomena such as spread-F, sporadic E, polar cap patches, intermediate and descending layers, and traveling ionospheric disturbances. Ionospheric disturbances can result from such processes as plasma instabilities, auroral precipitation, meteors, geomagnetic storms and substorms, thermospheric winds, and gravity waves. Physics-based models that account for the development and evolution of these disturbances are required to achieve a full predictive capability. After such models have been developed, they must be integrated into the large-scale ionosphere models, such that the large-scale models set the conditions under which the smaller-scale disturbances will develop. These coupled models will enable predictions about the likelihood of ionospheric disturbances at a given place and time.

A.2.12   Ionospheric Scintillation Models

A global climatological model of scintillation, WBMOD, is available. The model was initially developed from scintillation observations with the Wideband satellite and phase

A-23

screen theory of scattering. The initial model was found to be deficient because of poor temporal and spatial coverage. These limitations arose from the Sun-synchronous orbit of the Wideband satellite and the limited number of observing stations. The WBMOD model has recently been upgraded by infusing time-continuous equatorial scintillation data from geostationary satellites and scintillation data obtained from the HiLat and Polar Bear satellites in the auroral/polar cap region. The upgraded WBMOD (UWBMOD) model specifies scintillation for any radio wave propagation path between the ground and a satellite above 1000 km and at any frequency above 100 MHz. The inputs to the model are the day number, universal time, sunspot number, magnetic index, locations of the receiver and the satellite, frequency of satellite transmission, and phase detrend interval. The model outputs are the amplitude scintillation index, defined as the ratio of the standard deviation of intensity fluctuations to the average signal intensity; phase scintillation index, defined as the standard deviation of phase fluctuations over a specified detrend interval; and phase spectral strength and phase spectral slope, which define the phase structure of the scintillating signal.

UWBMOD is a climatological model and it fails to reproduce the extreme day-to-day variability of scintillation. As such, it is of limited use for real-time operational support and is useful only for long-term planning purposes. In order to improve its usefulness, the model needs to be driven by scintillation data from a network of stations. Because the frequency range over which intensity scintillation can be extrapolated is limited to the weak-scatter domain, multi-frequency data will be necessary. From the operational standpoint, the model needs to provide support to systems in the very high frequency (VHF) to GHz range of frequencies; 250 MHz transmissions from communications satellites and 1.2 to 1.6 GHz transmissions from GPS satellites may be exploited. The optimum number of sensors is dictated by the spatial dimensions of scintillation structures, their motions and lifetimes. The real-time, data-driven UWBMOD, developed in the interim, will provide a nowcasting capability for scintillations.

The later goal is to develop a physics-based model for the purpose of forecasting scintillation. Such a forecast model will be based on existing first-principle electron density models that incorporate: the physical processes leading to the formation of macroscale (several hundred kilometer) structures, including polar cap patches and equatorial plasma bubbles; further structuring to mesoscale and smaller scales (a few kilometers to tens of meters) by plasma instabilities; determination of the saturation amplitude of mesoscale structures; and radio wave scattering theory that computes scintillation for the propagation of signals from satellites through such turbulent media. The physics-based model should be developed in a phased manner. It needs to be focused on the equatorial region where scintillations are most severe. The next area should correspond to the polar region, and finally the focus should be shifted to the mid-latitude region. Plasma instability computations will require physical inputs. Critical needs are measurements of E-field components in the F-region, and plasma drift in the neutral frame of reference. Some of these sensors are currently available on the ground and onboard the DMSP satellites.

A.2.13   Neutral Atmosphere Models (Thermosphere and Mesosphere)

The appropriate modeling tool for the study of atmospheric density structure is the class of numerical TIEGCMs which can self-consistently calculate density perturbations and neutral wind systems on a global, 3D, time-dependent basis from physical principles.

As a first step in thermospheric nowcasting, empirical or semi-empirical models, exemplified by the Mass Spectrometer and Incoherent Scatter (MSIS) model, have been continuously improved and extended over the years. These models provide a critical first-order validation of any operational model and, indeed, can be used as such where high spatial resolution or time-dependency is not an issue. It is important to continue the development of these models that enable different data sets to be quantitatively reviewed and intercompared. Several operational models of thermospheric density and temperature, currently based on such semi-empirical models, reflect the mean behavior of the thermosphere as described by the large bodies of data from the various previous experimental programs. The semi-empirical or hybrid models use analytical functions to fit to randomly selected subsets of the available data, and spectral coefficients are generated that can be used conveniently to reconstitute thermospheric densities and temperatures as a function of space, time, and geomagnetic and solar activity levels.

The most important limitations of current numerical models are related to the accuracy of the time-dependent parameterizations used for the upper and lower boundary conditions. For example, the E-field distribution at high latitudes is a key factor in determining the magnitude of thermospheric response to geomagnetic activity. Three separate approaches are being used to lift these limitations. First, attempts are being made to lower the lower boundary of the TIEGCM to altitudes below the mesosphere to enable more self-consistent calculations of the important dynamical structures. Second, more rigor is being applied to calculations of the topside parameters (heat fluxes, currents, etc.), with the ultimate goal of generating a coupled magnetosphere-thermosphere-ionosphere model. Third, more systematic approaches are being used to develop improved empirical representations of boundaries, as in the assimilative mapping of ionospheric electrodynamics model, which uses disparate data sources to derive improved E-field distributions.

The fully coupled thermosphere-ionosphere models being developed for nowcasting and forecasting will use a variety of prescriptions and parameterizations based on geophysical and solar indices to describe the thermospheric energy inputs and solve the coupled governing equations discussed earlier to calculate all the thermospheric state variables (temperature, density, composition, and wind) on a global time-dependent grid. These models are being validated, and results are encouraging, particularly at F-region altitudes where Dynamics Explorer satellite data have been available to constrain and test the formulation.

A.3   Observations

Users of space weather information are concerned with both the background environment and the time of arrival, intensity, and duration of space weather disturbances as they manifest themselves at specific locations. The variations occur on time scales from minutes through days to years. To have a complete picture of the environment from the present into the future, forecasters need observations from key locations beginning with the origin of the disturbances at the Sun and continuing along their propagation routes into the near-Earth environment.

Eventually, a complete suite of observational sensors will provide this information on a continuing basis. However, in the near term, many sensors will be developed and deployed to provide observations that will support research into several areas: helping to improve the understanding of the physics of space weather, providing information to improve space sensors, and defining the requirements for observations to support operational models. This section describes the observations required to meet these needs.

A.3.1   Solar/Solar Wind Observations

CME. Investigations along the following lines should prove fruitful in improving our ability to predict CME-related disturbances in the solar wind and the associated energetic particle events:

- Develop techniques useful for understanding and predicting solar wind disturbances using soft x-ray images of the Sun such as provided by Yohkoh and the Solar and Heliospheric Observatory (SOHO), in anticipation of the Solar X-ray Imager (SXI) x-ray monitoring spacecraft series. This will enhance the x-ray observations currently being made by the Geostationary Operational Environmental Satellite (GOES).
- Study and assess radio capabilities for tracking solar wind disturbances in interplanetary space from the Sun to the Earth, including observations of both radio bursts and the interplanetary scintillation technique.
- Optimize coronagraph capabilities for studying the behavior of CMEs as a function of radial distance.
- Maintain ground-based coronagraph observation activities to provide a measure of global solar CME activity levels, and to increase the database for investigations of solar cycle variations of CMEs.
- Investigate the utility and capabilities of a solar wind monitor placed near Venus or Mercury orbit for predicting solar wind disturbances near Earth.
- Maintain an L1 or equivalent upstream monitoring capability for carrying out the above investigations and for maintaining at least a 1-hour forecast capability for major geomagnetic storms.
- Develop an EUV magnetograph capable of measuring coronal magnetic fields and test it on a spacecraft mission.

A-26

Solar Flare. Several missions can greatly improve our understanding of the flare process if we are able to take full advantage of them. In particular, SOHO provides line-of-sight magnetograms with much higher spatial resolution than previously available from the ground. In combination with Yohkoh, the SOHO observations greatly increase our understanding of magnetic structure in the corona. SOHO affords a unique opportunity to observe flares into the 2000 maximum from optical through soft x-ray wavelengths.

Technology developments are also under way that could constrain some of the models of flare energetics and improve our ability to predict the space weather consequences of an event on the Sun. In particular, the technology for imaging spectroscopy of hard x-rays with spatial resolution of less than 2 arc-seconds is now available. The current maximum provides an opportunity for applying this new technology to obtaining observations of many solar flares. Such observations could revolutionize our understanding of the basic problems of flare particle acceleration and heating.

Additional new technologies are poised to improve our understanding of the crucial magnetic field configuration in active regions and its relation to the larger scale magnetic field structure in the corona. The Flare Genesis project, consisting of an 80-cm solar telescope, was launched in December 1995 from the Antarctic. Flare Genesis was designed to obtain full-vector field observations above the atmosphere using a system of filters, thus leading to breakthroughs in our understanding of flare energy buildup and release. If Flare Genesis is successful in attaining its science objectives, balloon-borne vector magnetographs should be flown throughout the Solar Maximum.

These new missions and technologies notwithstanding, ground-based observation is the only likely option in the near future for continuous monitoring of the solar magnetic field. Present-day vector magnetographs are relatively insensitive; however, recent demonstrations of the use of infrared (IR) telescopes and magnetographs show their promise for obtaining measurements of the full-vector fields with much greater accuracy. This technology should be pursued vigorously and, if successful, developed for transition to an operational capability.

Existing routine observations can also be applied in new ways to the forecasting activity if the most recent understanding is exploited. For example, most flare forecasting is performed, at present, on the basis of the instantaneous structure of the photospheric and chromospheric magnetic, velocity, and photon emission intensity fields. By following the temporal evolution of these quantities, it should be possible to improve the accuracy of the predictions. For example, it may be possible to detect changes in magnetic shear leading up to a flare and to utilize those observations for forecasts. In addition, it is of special interest to predict flare intensity and duration. At present, this is done primarily by statistical tables that attempt to relate flare intensity and duration to other observable properties. These relationships need to be studied and tabulated much more accurately in order to decrease the high number of false alarms.

The observational goals pertaining to flare diagnosis and forecasting can be summarized as follows:

- Extend the SOHO and Yohkoh observations through the current solar maximum and expedite their utilization in models pertaining to flare-related space weather effects.
- Execute a flare mission for the current maximum. A mission such as NASA's High Energy Solar Spectroscopic Imager (HESSI) will utilize advanced high-resolution x-ray technology.
- Pursue the development of balloon-borne magnetographs like those in the Flare Genesis project, and assess their potential for operational use during the next solar maximum.
- Develop full-disk IR magnetograph technology for ground-based (and possibly space-based) monitoring of the full-vector solar magnetic fields.

Solar Radio Noise. Discrete solar radio bursts are detected and reported in real time by the Radio Solar Telescope Network (RSTN) operated by the Air Force and collocated with solar optical observing facilities worldwide. Plans call for RSTN to be replaced or augmented by the Solar Radio Burst Locator (SRBL), which will not only measure burst levels but also locate the solar position of the microwave source.

Data analysis and interpretation efforts, which employ new observations such as x-ray images from Yohkoh, are needed to investigate the nature of the sources of radio noise and their relationships with other features and events. Observational studies of the cause of radio noise should attempt to determine whether noise enhancements are precursors or byproducts of other significant releases of energy in the solar atmosphere. These studies could be undertaken immediately using currently available data.

Solar Wind. The numerical models of the solar wind need to be extensively compared with observations before they are incorporated into forecasting tools and tested operationally. Solar observations, especially of the photospheric magnetic fields, are required to provide the boundary conditions at the "source surface," although theoretical extrapolations or approximations are required to map these conditions to the ~20 R_S distance where the flow becomes more nearly radial and the speed is thought to be fully established. Currently, full-disk magnetograms are obtained routinely (e.g., daily) at observatories such as Wilcox Solar Observatory, Big Bear Observatory, Mt. Wilson Observatory, and the National Solar Observatory at Kitt Peak. Archives of interplanetary data at the National Space Science Data Center (NSSDC) have become widely available and could be used in conjunction with the magnetogram data to assess a given model's accuracy. The necessity of observing the photospheric fields more frequently as a part of model validation and application tests remains to be explored, as do other means of observationally determining the global solar wind inner boundary conditions. The time is ripe for experimentation with solar wind models, both to test them in retrospective data comparisons and to try predictions related to the generation of the primary "geoeffectiveness" parameters: velocity V, vector magnetic field B, and dynamic pressure ρv^2 that are observed at the L1 point by various spacecraft such as WIND and ACE. Ulysses observations currently provide a valuable 3D view of heliospheric magnetic field and solar wind structure, which will improve the source surface models.

From another perspective, interplanetary scintillation (IPS) measurements in principle provide a means by which coronal holes and stream interaction regions can be remotely sensed. As with their use in CME detection, these data require interpretation of a line-of-sight integration through the solar wind. Nevertheless, to the extent that these can be used to monitor the solar wind near its source, they are regularly available data that should not be overlooked. The IPS data should be considered as complementary to in situ plasma data, but of course interplanetary field perturbations can only be inferred. On the other hand, recent observations suggest that the heliospheric current sheet can be remotely detected using this method.

A number of actions can be undertaken in the observational area to enhance our solar wind knowledge and our use of it in space weather systems:

- Upstream monitors providing at least V, B, and solar wind density can be maintained to both test solar wind forecasting models in retrospective analyses such as those described above and to make ~1-hour forecasts and nowcasts.

- IPS data from appropriately located spacecraft should be made routinely available for analysis and operational use. International efforts to improve IPS techniques should be supported and the results evaluated.

- The full-disk magnetic field should continue to be regularly monitored to both validate models and attempt long-lead-time predictions of solar wind behavior using the models. In addition, methods should be developed to determine observationally the full-disk vector B field on a regular basis (e.g., with IR magnetographs), because this forms the basis for many of the potential forecasting model boundary conditions.

- Methods for determining the solar wind velocity near the Sun from remote observations need to be developed, including those based on full-disk magnetic field observations.

- The Solar Probe mission should be developed and used to learn more about the solar wind origin and acceleration, and its connection to the solar magnetic field structure.

A.3.2   Magnetospheric Observations

For operational nowcasting and forecasting, continuous, real-time data are crucial. Space-based measurements include upstream solar wind and IMF data; particle and field data from at least four well-positioned geosynchronous satellites; particle data from the suite of GPS satellites; and particle, magnetic field, electric field, and imaging data from low Earth orbit (LEO) polar orbiters. Currently, LEO operational satellites include both the DMSP and National Oceanic and Atmospheric Administration (NOAA) Polar Orbiting Environmental Satellite (POES) systems, which will eventually converge to form the joint National Polar Orbiting Environmental Satellite System (NPOESS). Assets currently in geosynchronous orbit include the GOES 8 and 10 satellites, carrying magnetometers and energetic particle instruments, and several satellites carrying Department of Energy environmental plasma and energetic particle instruments. Ground-based systems include instruments for observing surface magnetic field data that adequately cover low-, middle-, and high-latitude regions (for example, as described in The National Geomagnetic Initiative, published by the National Research Council, National Academy Press, Washington, D.C., 1993); the existing and planned radar arrays; and ionosonde and riometer stations. All these measurements have an important role to play in an effective operational forecasting capability.

Operational measurements must be available in real time if they are to be of use for prediction. Real-time data access will be important for both ground-based and satellite-based data sources. Operational space weather prediction of high-latitude phenomena requires an upstream monitor of solar wind conditions. The ACE satellite became the first operational monitor, and this must be followed by continuous operational monitoring of the IMF and solar wind parameters. All numerical space weather models ultimately require solar wind information to function in the predictive mode. The space environment sensors on the DMSP and NOAA POES satellites provide an especially valuable resource for constraining the present-generation models and will continue to be useful for future models as well. Improvement of the freshness of the data is necessary for predictive capabilities. Other fleets of operational satellites should be utilized in the future to provide critical data for constraining space weather models. One source of especially useful data could be the GPS constellation of satellites, which could provide data on the source population temperature and other moments of auroral particles prior to acceleration by field-aligned potentials, parameters of particular use in the modeling of magnetosphere-ionosphere coupling.

Observations for attacking the science issues and observations for providing operational drivers for space weather models are overlapping but different. During the period leading up to and through the solar maximum, a large number of research satellites have been and will be launched that can provide critical data for investigating key science issues. NASA's POLAR and WIND satellites, combined with the Japanese Geotail mission, will answer basic questions on a global scale of energy input, storage, and release relative to magnetic storms and substorms. The Fast Auroral Snapshot (FAST) satellite will investigate auroral processes on microscales and mesoscales. The Ørsted satellite will make very accurate measurements of the near-Earth magnetic field, which will be useful for the study of external currents. Put in the context of the auroral images from POLAR, these measurements will be used for addressing the 5-km-scale description of aurora desired by the customer base. Continuing these missions through the more active conditions typical of the solar maximum will ensure that the physics learned fits not only moderate to active conditions but also the extreme conditions that can be the most devastating for space weather-related problems. Continuous monitoring of the high-latitude regions by the DMSP and NOAA POES satellites and by ground-based techniques such as incoherent scatter radars, HF radars, magnetometers, and riometers provides complementary information to satellite missions and is an essential component of the research database needed to explore and define the physical processes. Extension of ground-based facilities to the center of the polar cap with the Relocatable Atmospheric Observatory augments the current longitudinal-chain measurement capability significantly. This complements the Super Dual Auroral Radar Network (SuperDARN) chain of HF radars that will monitor simultaneously auroral zone electrodynamics over 7 hours of local time. The Automated Geophysical Observatories (AGOs) will provide information about magnetic and auroral conditions in the Southern Hemisphere

An important future thrust for magnetospheric observations is the validation of techniques for space-based imaging of the magnetosphere. In addition to existing techniques for multispectral imaging of the aurora, the technology exists to image the ring current, plasmasphere, inner plasma sheet, and possibly ionospheric ions. Many of these concepts will be validated as part of NASA's Imager for Magnetopause to Aurora Global Exploration (IMAGE) satellite program. Other future research satellites of importance include a dedicated radiation belt satellite and a satellite in the IMP-J orbit to monitor the near-Earth solar wind and tail regions.

A.3.3   Ionosphere/Thermosphere Observations

Electron Density. The most critical need for the calculation of ionospheric electron densities is a specification of the time-dependent model inputs on a global scale (i.e., specification of the convection and precipitation patterns, meridional neutral winds, and dynamo E-fields). Also, for validation purposes, global measurements of the electron density distribution are required. The inputs and electron densities are needed for a range of geophysical conditions, including quiet and disturbed periods for different seasonal and solar cycle conditions. Ultimately, multi-site measurements from a globally distributed network of relatively inexpensive instruments operating in real time are required for forecasting purposes. Coordinated multi-instrument measurement campaigns will also be needed to address unresolved physics issues.

In the near term, progress in the physical understanding of coupling processes and time delay mechanisms can be achieved by exploiting the comprehensive databases recently acquired as part of Coupling Energetics and Dynamics of Atmospheric Regions (CEDAR) and Geospace Environment Modeling (GEM) campaigns. These databases include multi-site measurements and data from a variety of instruments (magnetometers, coherent and incoherent scatter radars, satellites). Because the largest perturbations and the greatest uncertainties in density calculations occur during geomagnetic storms and substorms, the near-term observational emphasis should be on these phenomena. However, there are still important unresolved issues connected with the creation, transport, and decay of electron density structures, and multi-instrument measurement campaigns are required to resolve them. In addition, it seems certain that more LEO receivers on spacecraft like GPS-Met will be launched for meteorological purposes, but will also be potentially very useful for determining ionospheric vertical density profiles by the use of inversion algorithms.

Ionospheric Electric Field. For advancement at high latitudes, the observational requirements are clear. We require the maximum possible coverage of the high-latitude region with specification of the F-region ion drift or electric field. This need is met in part with radar arrays that may be augmented in the near future. In this area there is a real synergism between the capabilities of ground-based arrays to provide good temporal resolution with less than global coverage and spacecraft systems to provide almost global "snapshots" at given times. An ideal satellite configuration, for adequate constraint of the models, places sensors in at least two and preferably four local time planes with multiple spacecraft in each plane. Such a configuration, together with ground-based sensors, would dramatically advance our ability to determine the nature of the temporal and spatial distribution of the potential, and to begin a meaningful parameterization of its evolutionary properties.

In the long term, the Relocatable Atmospheric Observatory (RAO) will provide invaluable data for resolving problems that have plagued our community for almost 30 years. For example, although E-fields have been measured since the early 1960s, there is still a major disagreement on the configuration of the high-latitude convection pattern when the interplanetary magnetic field is northward. The combination of data from the RAO and the DMSP satellites should resolve this issue.

At middle and low latitudes the need for monitoring the **E×B** drift of the plasma is paramount. The spatial scales of interest are such that this could be rather easily accomplished with a low-inclination spacecraft. The use of ground-based measurements to describe the height of the F-peak will continue to make important contributions in this area. However, access to a data set from a satellite in equatorial orbit would allow the association between **E×B** drift motion and the nighttime stability of the ionosphere to be quantitatively established.

Ionospheric Disturbances. Among the requirements is global, long-term (10-30 day), around-the-clock observations that by their very definition are guaranteed to capture quiet and storm-time conditions and associated transitions in the growth and recovery phases of associated disturbances and their inter-hemispheric manifestations. This will permit an accurate empirical specification of the weather and climatology of E, F_1, and F_2 characteristics. In this regard, archives exist with sufficient campaign data (e.g., in the CEDAR database) to carry out a systematic quantitative evaluation of empirical and first-principle models and determine levels of accuracy and inadequacies in model prediction capabilities as a function of season, solar epoch, local and universal time, geographic and geomagnetic domains, and levels of disturbance. With this perspective, the existing ionosonde database (with more than 50 stations currently active worldwide) can provide heights and densities of the E- and F-regions along with specification of dominant intermediate and descending layers. The F-region densities will define the temporally and spatially dependent condition of the ionosphere in its quiet, transitional, and disturbed states, and the measurements of F-region heights, through servo-analysis, will provide an indirect observation of meridional winds at all local times over a broad range of mid-latitude sites. Ionosondes also provide direct measures of sporadic-E as well as range and frequency of spread-F. Augmentation of these observations with incoherent scatter radars (for E-fields and F-region winds), Fabry-Perot interferometers (for dawn/dusk F-region winds), and satellites of opportunity, e.g., DMSP and NOAA POES for specification of high-latitude inputs, will provide significant gains. Future programs can build on more complete specification of the ionosphere-thermosphere system and controlling forces. Promising opportunities include the Arizona Airglow Experiment (GLO), Midcourse Space Experiment (MSX), Advanced Research and Global Observing Satellite (ARGOS), and Thermosphere-Ionosphere-Magnetosphere Energetics and Dynamics Mission (TIMED) programs and a recently proposed upper F-region satellite at low inclination focused on topside profiles and scintillation data during conditions of equatorial spread-F.

Ionospheric Scintillations. Effects of scintillation on some operational systems, notably GPS satellite navigation systems operating at 1.2-1.6 GHz frequencies have not been determined. It is generally recognized that GPS navigation systems are vulnerable in the polar and especially in the equatorial region during the solar maximum period. In the equatorial region the irregularity structures are highly elongated in the north-south direction and are discrete in the east-west direction with dimensions of several hundred kilometers. With such spatial distribution of irregularities, we need to determine how often a GPS receiver fails to provide navigation aid with the available constellation of GPS satellites. It is recommended that GPS receivers acquiring amplitude and differential phase at a high data rate (50 Hz) be deployed to assess the effects of scintillation on the performance of GPS navigation systems in the equatorial region. Such measurements will also be helpful in the design of satellite-based cellular telephone systems using L-band frequencies.

Observations are also needed to establish the survivability of MILSTAR satellite systems using frequencies in the K_a band (20-40 GHz) and the UHF band (250 MHz). K_a-band transmissions are not affected by the ionosphere but are very vulnerable to rain and fog in the equatorial region. On the other hand, 250 MHz systems will be able to withstand the effects of rain but suffer outages from ionospheric scintillation.

In view of the extreme spatial and temporal variability of scintillation, immediate deployment of a network of ground-based scintillation sensors is recommended. Sensors need to cover the frequency range (250 MHz to 6 GHz) used by operational systems. Geostationary satellite transmissions of 250 MHz and GPS satellite transmissions at 1.2-1.6 GHz may be utilized by these sensors. The current climatological model of scintillation may be adapted to ingest the multifrequency scintillation data to provide a real-time-data-driven scintillation model.

From an operational standpoint, the forecasting and specification of equatorial scintillation is a major requirement. From a physics point of view, the temporal and spatial variability of scintillation in the equatorial region remains unresolved. There is general agreement that the interaction between thermospheric neutral wind and the ionized species critically controls the E-field at F-region heights in the post-sunset period, and thereby controls the onset of plasma instabilities and scintillation. Whether the instability is inhibited by the meridional wind on a day-to-day basis or it requires a finite-amplitude seed perturbation from gravity waves is not known. Overall, there is a need to monitor the thermosphere-ionosphere interactions on a short temporal and spatial scale. An equatorial satellite with altitude between 600 and 700 km and orbital inclination of about 30° carrying a modest number of sensors is required to monitor this interaction. The sensors should include an ion drift meter, vector electric field instrument, phase coherent radio beacons at L-band and VHF frequencies, and instruments to measure neutral wind and ion/neutral composition. With an orbital period of 90 minutes, the satellite will be able to monitor the formation of plasma bubbles and their scintillation effects on beacon transmissions. It will also support studies on thermosphere-ionosphere interactions at low latitudes and provide physical inputs to the physics-based model of scintillation.

Neutral Atmosphere. The neutral upper atmosphere is a data-poor domain by comparison with the lower atmosphere. In fact, the existing database is still insufficient to provide a full

A-33

climatology of thermospheric state parameters as a function of altitude, universal time, local time, latitude, longitude, season, solar activity and geomagnetic activity. Important pieces of this climatology certainly exist, but careful experimental work is still needed to provide a complete quantitative description. Semi-empirical models, such as MSIS, may be used for rough estimates of thermospheric parameters. The accuracy of such models, however, is still quite limited; for example, mean thermospheric densities are modeled to accuracies of only 15% at best, with errors of factors of two or more occurring at high latitudes and/or for geomagnetically active periods. Some of the problems can be traced to spatial- and temporal-resolution limitations, some to instrument calibration limitations, and some to limitations in spatial coverage. Clearly, these models will need to be greatly improved through the ingestion of new, carefully calibrated, global-scale data sets. Similar considerations apply for the numerical, hybrid, or stripped-down numerical codes, which all require more extensive, well-calibrated global-scale data sets for validation and testing. The basic observational needs are for measurements of thermospheric and upper mesospheric density, wind, temperature, and composition to accuracies of better than 5%. Careful attention to spatial and temporal coverage will be required to aid in the specification of the thermospheric and mesospheric waves of importance (gravity waves, planetary waves, and tides). This latter challenge will require the systematic deployment of airglow imagers, lidars, HF radars, and specialized satellite techniques.

The thermosphere is driven by solar EUV heating. Currently, however, the ground-based proxy indices for solar radiative fluxes are inadequate. Therefore, it will be of central importance to develop a long-term, reliable monitoring method to measure the direct solar EUV fluxes impinging on the upper atmosphere. This will require a long-term commitment to deploy a satellite instrument or series of instruments. Also, sustained observations of thermospheric structures at high geomagnetic latitudes will be needed from satellites, the RAO, and other sites, to determine the nature of the inputs and atmospheric response to geomagnetic activity. This effort would be greatly helped by the recommended high-inclination, highly elliptical satellite designed to investigate high-latitude electrodynamics.

Finally, the development of new experimental means of monitoring thermospheric compositional changes from ground and space will be very advantageous. It is clear that a major limitation of current thermospheric and ionospheric models relates to the ability to predict composition of such critical species as atomic oxygen, nitric oxide, and carbon dioxide.

APPENDIX B

NATIONAL AERONAUTICS AND SPACE ADMINISTRATION LIVING WITH A STAR

B.1 Introduction

Sun-Earth Connections is one of four major themes of emphasis within the National Aeronautics and Space Administration programs. One of the four key questions to be addressed under Sun-Earth Connections is how solar variability affects life and society. At the end of 1999, NASA put forward its Living With a Star (LWS) Program to advance scientific understanding of solar variability and its effects. LWS has four main objectives:

- Identify and understand variable sources of mass and energy coming from the sun that cause changes in our environment with societal consequences, including the habitability of Earth, use of technology, and the exploration of space.
- Identify and understand the reactions of geospace regions whose variability has societal consequences (impacts).
- Quantitatively connect and model variations in the energy sources and reactions to enable an ultimate US forecasting capability on multiple time scales.
- Extend the knowledge and understanding gained in this program to explore extreme solar terrestrial environments and implications for life and habitability beyond Earth.

Although the objectives of LWS are broader in scope than the National Space Weather Program, many of the planned activities support space weather goals. LWS will provide the scientific context and understanding that will lead to a capability to accurately predict solar activity and its effect on the space environment. Initially the main application of such knowledge is to geospace, but as mankind starts to explore the solar system, the requirement is extended to interplanetary space and other planets, specifically Mars.

LWS seeks solutions to a broad class of problems associated with the dynamic effects of solar, interplanetary, magnetospheric, and upper atmospheric phenomena that have an impact on humans. In particular, it is focused on the impact of these phenomena on modern technology and on the safety of humans as they travel beyond the Earth.

LWS is a cross-cutting initiative whose goals and objectives have the following links to each of the four NASA Strategic Enterprises:

- **Space Science** - LWS quantifies the physics, dynamics, and behavior of the Sun-Earth system over the 11-year solar cycle.
- **Earth Science** - LWS improves understanding of the effects of solar variability and disturbances on terrestrial climate change.
- **Human Exploration and Development** - LWS provides advanced warning of solar energetic particle showers that affect the safety of humans in space.

- **Aeronautics and Space Transportation** - LWS provides detailed characterization of radiation environments useful in the design of more reliable electronic components for air and space transportation systems.

B.2   LWS Program Elements

The following subsections describe the LWS strategy for studying the effects of solar variability.

B.2.1   Accelerate Solar Terrestrial Probes

Studying the Sun-Earth connected system requires simultaneous observation of interacting regions. In the current budget, missions have two-year design life and are launched at 2.5-year intervals. This limits synergism between missions studying different regions of the Sun-Earth system. The LWS goal is an interval of 1.5 years between missions to enable simultaneous study of key linked regions in the Sun-Earth system.

In addition to currently planned missions, LWS adds observational and modeling capabilities to fill gaps in coverage of key regions of the Sun-Earth system. The gaps and the planned solutions are listed below.

Missing: Detailed information on dynamics of the solar interior and the dynamo that generate and control solar variability *(including both short and long term variations)*.

Solution: A Solar Dynamics Observatory providing high time and spatial resolution data to probe the:
- Solar interior and the subsurface structures underlying regions generating solar disturbances.
- Dynamics of magnetic structures in the solar atmosphere where these disturbances occur.

Missing: Continuous observations of solar regions generating solar disturbances; measurements of the solar interior from the other side of Sun.

Solution: Solar Sentinels:
- To observe the entire solar surface, *including the far side from Earth.*
- To observe, globally and in stereo, solar wind disturbances from the Sun into interplanetary space.
- To obtain "missing" seismology data from the solar far side.

Missing: Detailed information on the dynamics of the terrestrial space environment during geospace disturbances.

Solution: Geospace Dynamics Network: A network of spacecraft to provide data with sufficient spatial and temporal coverage to specify the dynamics of disturbances affecting geospace and the neutral atmosphere.

Missing: Detailed information on long term effects/relation to global change of variable solar inputs on upper terrestrial atmosphere.

Solution: Geospace Dynamics Network

Missing: Theoretical understanding and end-to-end modeling capability ranging from solar sources to effects in geospace and the upper atmosphere.

Solution: Dedicated theory and modeling program

- To understand the physical connection between the solar source, geospace and the upper atmosphere.
- To specify space environment parameters of relevance to humans and human assets.
- To enable a predictive capability of the impact of solar disturbances on humans and human assets.

B.2.2 Establish a Space Weather Research Network

The program proposes a Space Weather Research Network (SWRN) comprising a distributed network of spacecraft providing continuous observations of the Sun-Earth system. The network would contain two key components:

- The *Solar Dynamics Network* would observe the Sun and track disturbances from the Sun to the Earth.
- The *Geospace Dynamics Network* would comprise constellations of small satellites in key regions of geospace to specify the dynamics of disturbances affecting geospace and the neutral atmosphere.

The SWRN is shown in Figure B-1 and described in more detail below.

B.2.3 Establish Targeted Data Analysis and Modeling

LWS can exploit data from present and past missions:
- To improve knowledge of space environmental conditions and variations over the solar cycle.
    - Obtain reliable environmental specifications for cost-effective design of spacecraft and subsystems to minimize space environmental effects and damage.
    - Important for commercial satellites and military space systems which must have "all weather" capability.
- To develop new techniques and models for predicting solar/geospace disturbances which affect human technology. For example, recent research has revealed a connection between sigmoidal x-ray signatures and regions with high probabilities for producing CMEs.
- To develop cost-effective techniques for assimilating data from networks of spacecraft.

B.2.4   Establish Orbital Technology Testbeds

LWS provides the opportunity for low cost validation of radiation-hard and radiation-tolerant systems in high radiation orbits. This opportunity can be exploited through NASA, industry, DOD, and other agency partnering.

B.2.5   Establish and Expand Partnerships

- Major contribution to the National Space Weather Program (NSWP).
- Establish/expand collaborative research efforts with DOD, DOE, NOAA, NSF.
- Coordinated research and development on the space environment and space weather.
- Launch of LWS missions as secondary payloads on Evolved Expendable Launch Vehicles (EELVs).
- Use of LWS real-time/near real-time data for operational purposes.
- Establish partnerships for flight of NASA- provided and/or designed space environmental sensors on commercial and other government agency spacecraft.

B.3   The Space Weather Research Network

The Space Weather Research Network (SWRN) would provide crucial measurements from a series of critical vantage points distributed around the Sun and Earth. See Figure B-1. The results would then be brought together and analyzed in a coordinated fashion to discover the essential knowledge, to provide warning of specific events that will affect our space assets, and to relieve effects on our increasingly technology-reliant society.

The SWRN draws heavily upon other NASA missions, although their implementation may be modified if they are applied solely to space weather investigations. The ultimate output of the program would be the scientific knowledge and observational specifications to improve operational space weather systems, and the models to apply to the data to produce accurate and reliable forecasts over the time-scales required to be beneficial to humanity's space endeavors.

B.3.1   A Global View of the Sun

The first part of SWRN is to obtain the required measurements of the Sun, the driver of the system. The output of particles and fields in the form of the solar wind and radiation across the whole electromagnetic spectrum can affect our environment in many different and specific ways.

It is vital to observe the entire Sun, including the far side, which is not visible from Earth. For example, from a medium-term (days) prediction point of view, it is important to know whether an active region about to rotate onto the solar disk has grown or decayed while on the far side of the Sun. It is then possible to better assess its likelihood of producing a major flare or CME. Events, such as coronal mass ejections, on the far side of the Sun can affect space weather.

From a longer-term prediction (weeks) point of view, it is important to be able to "see" how the sub-surface magnetic field is evolving before it erupts into new sunspot groups. This is now possible with the new science of helioseismology.

Predictions of the next solar cycle (years) would require continuous observations of the solar polar regions where the reversal of the global solar dipole first becomes evident. Also precise measurements of the solar irradiance as a function of wavelength is required to determine the solar input to the Earth and planets. Similarly, measuring the amount of energy lost by the Earth across the spectrum is equally vital.

The specific measurements we need to obtain a high-fidelity view of the Sun are:

- Remote sensing of the internal dynamics of the Sun from a geosynchronous vantage point (Solar Dynamics Platform) which provides a capability of high data rate. This mission could be combined with the irradiance measurements of the Sun and Earth (i.e., an extended version of SONAR).
- Remote sensing of solar activity from key vantage points around the ecliptic (i.e., a next generation STEREO mission including a Solar Farside Observer component).
- Remote sensing of the solar polar regions (i.e., a version of the Solar Polar Imager mission)

Such a group of "Sentinel" missions would produce the data we require individually, but operating together they would provide us with a powerful predictive tool for space weather. Assuming they were designed in a coordinated way, they could also make true tomographic (3D) images of solar events.

While such measurements would characterize solar activity on all the required time-scales in a coordinated fashion for the first time, we also need to know the effects that our ever changing Sun has on the inner heliosphere and understand how those perturbations propagate out from the Sun towards the Earth.

B.3.2 Transit of the Solar Wind

The second stage of the SWRN program is to see how solar disturbances propagate towards the Earth. Their passage through the interplanetary medium dominated by the solar wind changes their nature and effect on the planets they encounter. It is vital that we use both in situ and remote sensing techniques to sample their progress and evolution.

Such measurements also give us a short-term alert (hours) to the approach of a solar disturbance, such as a CME. The required measurements consist typically of particle, fields and plasma wave instruments that characterize the composition, velocity and density of the solar wind as well as the strength and direction of the imbedded magnetic field. This vector information determines the geoeffectiveness of a given event. The transit of CME mass will be tracked with remote sensing techniques (e.g., Stereo CME imaging).

Such measurements should be made from the solar "Sentinel" missions discussed above. There needs to be a small group of such sensors "upwind" from the Earth, within the solar wind stream that will impact the Earth. The L1 point has been ideal for such a group of instruments to characterize the global as well as the local properties of the solar wind disturbances that may cause strong geomagnetic storms.

B.3.3   The Global Properties of Geospace

The previous two parts of the SWRN program provide warning and characterization of the events as they approach the Earth. The last and most complex part of the puzzle is determining how a given event will impact geospace.

To do this we must understand how the energy "leaks" through our magnetospheric shield, how it is redistributed and dissipated. These are global phenomena that are often controlled by local microphysical processes. Hence we have to use a combination of remote sensing to sample a broad range of the macroscale phenomena and widely distributed in situ measurement techniques similar to those used in the heliosphere to understand the microscale processes.

Hence the measurements we need to make are:
- Remote sensing of the Earth's polar regions and night side to see auroral development (i.e., an extension of the Pole Sitter concept to include both poles and the Lagrangian location.)
- In situ plasma and field measurements of the Earth's radiation belts (Radiation Belt Mappers) to see how they are affected by such events (i.e., a version of IMC)
- A combination of remote sensing and in situ sampling of the Earth's upper atmospheric layers (Ionospheric Platforms) to see how these effects propagate down into inner geospace and how they affect the Earth itself

B.4   A Coordinated Approach

This exciting program would be the first step to creating a comprehensive space weather forecasting capability. However it also requires that the data be brought together in a systematic and coordinated fashion, and that LWS missions are supported by a solid theory and modeling program. Only in this way can we study the Sun-Earth connection as a system. A new breed of interdisciplinary scientist must be encouraged to approach the space weather problem in much the same way as we solved the problem of inaccurate short-term meteorological forecasts 25 years ago.

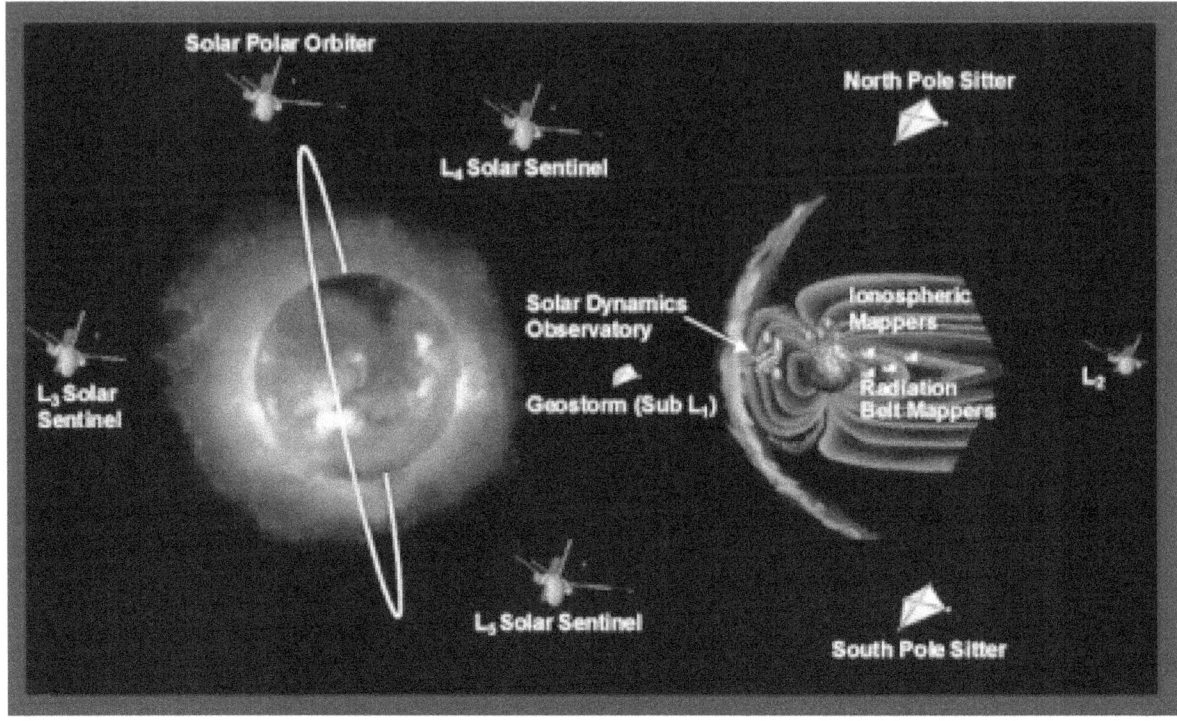

Figure B-1. Space Weather Research Network

APPENDIX C

NATIONAL SECURITY SPACE ARCHITECT
SPACE WEATHER ARCHITECTURE

C.1   Introduction

The Space Weather Architecture was developed using the NSSA standard architecture study development process. The Space Weather Architecture Study Terms of Reference (TOR), 4 December 1997, directed the NSSA to lead an integrated Space Weather Architecture Study with Department of Defense (DOD), National Aeronautics and Space Administration (NASA), National Oceanographic and Atmospheric Administration (NOAA), and other agency participation. Accordingly, the NSSA formed a Space Weather ADT composed of representatives from major stakeholders. The ADT assessed the current environment and projected one 15-25 years in the future to understand how space capabilities may be used.

The Space Weather Architecture Study was conducted in two phases. Phase I determined that an architecture study was warranted and gathered the information necessary to conduct it. Phase II developed and analyzed architecture alternatives, and generated space weather architecture findings and recommendations. These key study findings and recommendations are listed below. The NSSA study results are consistent with the NSWP recommendations and reflect a more in-depth review of the 2010+ user needs, support of national priorities, and consider fiscal resource limitations.

After study completion, the NSS SSG endorsed an Architecture Guidance Memorandum that identified the Office of the Assistant Secretary of Defense for Command, Control, Communications, and Intelligence (OASD(C3I)), in coordination with NOAA, as the overall agency responsible for overseeing a Space Weather Transition Team, composed of key space weather stakeholders. A Space Weather Transition Team was organized to develop a plan to provide guidance on implementing the approved recommendations. The NSS SSG also provided direction in an Architecture Implementation Memorandum on implementing the Space Weather Transition Plan.

Periodic progress reviews of the Space Weather Architecture Transition Plan will be required over the coming years. The interagency coordinating structure of the NSWP Council and the CSW will provide the vehicle to assess the state of progress, adjust the direction of the program to match breakthroughs in technology or changes in requirements, and to resolve conflicts between or among the stakeholder agencies.

C.2   Key Study Findings and Recommendations

C.2.1 Space Weather Architecture Vector

To guide future investment, development and acquisition of space and space-related capabilities, the NSSA recommends:

- **Increase emphasis on Operational Model development**
- **Ensure improved Operational Capabilities based on User Needs**
  - **National Security priorities include Ionospheric and Radiation Environment Specifications and Forecasts**
  - **Civil priorities also include Geomagnetic Warnings and Forecasts**
- **Evolve to improved Forecast Capabilities, as phenomenology is better understood, models mature, and user needs are better defined**

Future National Security operations will require improved capability to accurately locate targets, provide precision navigation, and provide reliable mobile communications in a more time-constrained environment. To support these capabilities, immediate emphasis must be given to the accurate specification and forecasting of ionospheric total electron content (TEC) and scintillation parameters. It is essential that ground-based and space-based ionospheric observing systems and ionospheric models be developed and employed expeditiously. Also, significant to National Security is the capability to determine rapidly whether space weather or an adversary is degrading critical satellites. In addition, it is important to design robust satellites and rapidly recover damaged satellites. To support these needs, it is necessary to develop and employ systems and models to provide an essential capability to specify the radiation environment at satellite altitudes. The desired capability also includes forecasting of the radiation environment at satellite altitudes.

Like National Security needs, civil needs also include ionospheric and radiation specification and forecast. In addition, the civil community places priority on geomagnetic forecasting through in-situ solar wind measurements. These capabilities will significantly enhance the civil community's capability to support the power industry mitigation of losses from outages, and NASA's ability to protect astronauts from harmful radiation effects.

These recommendations are based on analysis of architectural alternatives. Because of the criticality of models to improved space weather architecture performance, these architectural alternatives have a common set of forecast and specification models. In addition, a common concept of operations (CONOPS) and communications systems were defined. Each alternative was designed to provide maximum benefit-cost ratio at its projected funding level.

One alternative represents a minimal capability. The sensor emphasis was on solar flare and Coronal Mass Ejection (CME) imaging and on providing a dense measurement grid of ionospheric properties and magnetospheric particles and fields. This alternative provides high quality specifications in all domains of interest such as ionospheric electron density, equatorial scintillation and neutral density, and solar event forecasts.

The second and target level, adds magnetic field and particle sensors near the L_1 position to directly sample the solar wind and the CMEs about one hour prior to their hitting the magnetosphere. The L_1 sensor adds confidence to ionospheric electron density, magnetospheric particles and fields and neutral density specifications, and improves polar scintillation, magnetospheric particles and fields forecasts.

The next level of development represents a more desired robust capability adding a second CME imager. It provides a more side-on view to better characterize evolution and minimize false alarms. It will improve longer-term forecasts of magnetospheric particles and fields, Van Allen Belt radiation, ionospheric properties, and neutral density.

Even this alternative is not expected to satisfy all the 2010-2025 needs, particularly forecast needs. Considering the likely technology and basic research and development necessary for space weather models implementation, this alternative is the best we can likely achieve during this timeframe.

These architecture concepts identified and pointed to the investments which would yield the most timely benefit. Specifically, the desired goal of achieving the capabilities represented by the most enhanced alternative must evolve from the minimal capability through the first level of enhancement as resources allow. To provide these capabilities, an integrated systems acquisition approach (e.g., sensors, processing systems, models, and products) focused on user needs is required.

Fiscal constraints demand prioritization of expenditures. The study identified the need for some increased investments. The highest leverage near-term investment was found to be validated, reliable space weather operational models. To achieve these models, a robust focused R&D effort is needed, including the continuation of science missions to collect data required by researchers for developing model algorithms. Operational data collection must provide increased sensor coverage and be archived to serve as a basis for model validation. This archive should be expanded to include correlated space weather impacts supporting system acquisition, simulation and operational planning.

This recommended architecture vector provides a user oriented approach, consistent with the OFCM plans. It will allow a smooth transition to the future national space weather infrastructure.

In summary, the ADT found:

- The Current Baseline supports limited model-development
- Primary space weather support future user systems needs are: Improved Ionospheric TEC & Scintillation and Radiation Environment specifications and forecasts
- Users desire continuously updated impact oriented products
- Increased investment in and dependence on space systems (military, civil, and commercial) justifies some increased space weather investments

C.2.2 Space Weather Importance Awareness

To guide future investment, development and acquisition of space and space-related capabilities, the NSSA recommends:

- **Integrate Space Weather information (system impacts and space weather environment data) into User Systems through inclusion in:**
  - **User Education**
  - **Simulations**
  - **Wargaming and Training**
  - **CONOPS**
  - **Contingency Planning**
  - **System Anomaly Resolution**
  - **Damage Assessment and Reporting**

National Security dependence on space support is increasing dramatically but the number of National Security satellites is expected to remain relatively constant with less backup and residual capability. Civil and commercial dependence on space systems is also increasing. Under these circumstances, each satellite is more critical and satellite outages will have greater impacts. The National Security demand for commercial SATCOM (e.g., hand-held terminals) will increase, creating new unpredictable vulnerabilities.

The study found limited design information and guidelines for many new orbits. Furthermore, commercial competitive pressures to cut satellite development costs lead to reduced testing and away from military hardening. The developers of user systems must be aware of potential space weather impacts through user education and space weather inclusion in Simulation Based Acquisition (SBA).

In the past, space systems have been part of the support (force enhancement) infrastructure. In the future, terrestrial weapons are likely to be directly targeted using space, and some weapons may be space-based. This will drive an increase in coverage, timeliness, accuracy, and command and control assuredness requirements for space systems. Space weather can significantly impact the ability to achieve needed levels of capability. For example, ionospheric scintillation can disrupt access to the Global Positioning System (GPS) and to radar signals with uncertainty in the ionospheric electron density degrading geolocation accuracy. Today, outage causes are often not precisely determined, leading to less effective mitigation, and recovery. This lack of understanding also impacts user system and space weather model design improvements.

Space weather fidelity in Service wargaming is seriously deficient. In essence, space weather is ignored. Better simulation of space weather effects in wargaming will increase space weather awareness in the user community and allow for development of mitigation and exploitation strategies.

In summary, the ADT found:

- Operators frequently do not understand space weather impacts. Consequently to reduce operational risks, space weather education and training is critical

- Space weather information needs to be integrated into all phases of system life cycles

- Budget and competitive pressures on satellite providers coupled with expected increases in demand for improved coverage, timeliness, accuracy and assuredness for space-based services, increase the potential impact of future space weather perturbations

- Space weather effects have the most impact on communications, position finding, navigation, timing, intelligence, surveillance, and reconnaissance

C.2.3 Space Weather Requirements

To guide future investment, development and acquisition of space and space-related capabilities, the NSSA recommends:

- **Develop a set of Approved Validated Space Weather Requirements focused on User Needs**
- **Update Requirements as User Needs and Technology evolve**

An effective space weather architecture depends on better understanding and documentation of user needs to provide compelling justification of what is needed and at what priority. Needs definition for the space weather study started with the draft AFSPC space weather CRD and the OFCM NSWP Strategic and Implementation Plans. Joint Vision 2010 was reviewed and its implementing systems evaluated for space weather susceptibility. Further understanding came from review of current architecture requirements and projected needs for a wide range of users. In addition, the User Applications Tiger Team systematically reviewed all classes of users (i.e., National Security, civil, and commercial) and their projected space weather impacts and product needs.

A space weather space architecture exercise (SAX) captured operators and planners opinions of user needs and potential user responses. This exercise examined the needs for product user interface, timeliness, and accuracy in several user system scenarios with space weather impacts. The results confirmed the need for a clearer definition of the requirements for the space weather architecture in user impact terms.

National Security users have a driving need for improved product confidence, accuracy, resolution, and coverage. Observations and measurement refresh rates must be increased to improve timeliness. Enhanced modeling and analysis techniques and rigorous validation will elevate forecast confidence. Improvement of product timeliness requires an enhanced capability to receive, process, and display space weather information.

In summary, the ADT found:

- Current requirements for space weather products are outdated, fragmented and incomplete

- Military and civil space weather requirements are similar but often addressed independently

- Insufficient understanding of user priorities and requirements causes significant gaps in current capabilities and has hampered efficient acquisition

- Space weather effects need to be translated into user impacts and evaluated for potential mitigation techniques

- Lack of users understanding of space weather impacts on operations has impeded development of accurate space weather requirements

- Requirements must be revised as user needs and technology evolve

C.2.4   Coordinated Space Weather Architecture Acquisition

To guide future investment, development and acquisition of space and space-related capabilities, the NSSA recommends:

- **Identify a cognizant organization in DOD to:**
  - **Manage the Acquisition of DOD Operational Space Weather Architecture and focus DOD Space Weather Research and Development**
- **Ensure Validated Models are developed in conjunction with Sensors and User Needs**
- **Ensure effective transitioning of R&D into Operations**
  - **Coordinate Acquisition and Integration of Space Weather Resources across Civil agencies and National Security Interest**

Needs for higher confidence user-friendly products are expected to grow. Model improvement is essential to increased specification and forecast confidence and performance.

Currently, the operational models are acquired from multiple sources—directly from R&D labs and universities, commercially, and through acquisition organizations. They supply products with differing or unknown levels of confidence. Sometimes the models have not been validated before quasi-operational implementation.

Historically space weather sensors were often fielded independent of the operational models or were not a-priori designed to work with operational models. A coherent user needs focus will lead to an improved space weather architecture performance.

The longest lead items for the architectures were found to be the models. This study identified and traced models to needed sensor inputs. It appears that efficiencies can be achieved by coordinating development of models between civil and National Security sectors. Within DOD, a single acquisition manager for the DOD portions of the space weather architecture can be achieved. To be most effective, space weather acquisition coordination needs to be performed at an interagency level.

In summary, the ADT Found:

- Military, civil, commercial and international cooperation will provide opportunities for cost and data sharing

- Cross-agency coordination is required to achieve improved model performance with validated model development prioritized to keep pace with sensor development

- Operational models, sensors and products can benefit from an integrated development approach

- The lack of a controlled process for model development and validation has led to inconsistencies in performance and confidence of models

C.2.5 Space Weather Information Archive

To guide future investment, development and acquisition of space and space-related capabilities, the NSSA recommends:

- **Consolidate and Expand the Existing Archival System**
  - **Capture Space Weather Environmental Data and System Impacts**
- **The Archival System should be:**
  - **Centrally Managed**
  - **User Focused**
  - **Incorporate Standard Formats**
  - **Accommodate Multi-level Security**

Spacecraft developers, insurance agencies, HF communicators, third party vendors, and power companies in the commercial sector responded to the space weather Architecture Study Request For Information. They provided insights into the needs for improved archiving.

(a) The commercial satellite builders are interested in historical space weather information (e.g., high, low, and average environments) to improve future satellite designs.

(b) Industry knows that design lessons are often relearned due to the long eleven year solar cycle and personnel turnover. Industry is increasing the pressure for reduced satellite development time and decreased testing time.

(c) Insurance rates currently do not reflect the space weather robustness of satellite systems, but interest was expressed in knowing the statistics of space weather events and impacts.

During the definition of user needs, the ADT determined that a significant number of space weather products must have high confidence. This requires that the models used to produce products be validated against the real world (i.e., historical space weather data (climatology) from multiple solar cycles and global coverage).

The difficulties of collecting validated space weather impacts on operational systems during Phase I pointed to the need for a centrally managed and standardized repository to capture impact information. Space weather impacts are often misidentified as other types of anomalies, increasing diagnosis time and thus the time to mitigate. These needs may be met by a centralized user-friendly data resource for researchers, space weather model developers, user system designers, planners and wargamers. It should capture space weather effects (start time, duration, and intensity), space weather climatology, and user system impacts.

In summary, the ADT found:

- Space weather effects and their operational impacts are not well documented— improved archiving of both would benefit research, operations, acquisition, analysis, simulation and wargaming

- Data that can be used to validate models and products is key to producing high confidence products

- Industry is interested in space weather design guidelines built on space weather climatological data

- Space weather impacts and environmental data are essential to understanding space weather trends

C.2.6 Integrated User Information

To guide future investment, development and acquisition of space and space-related space weather capabilities, the NSSA recommends:

- **Provide Space Weather Information:**
  - **In User Impact Terms**
  - **Routinely Available through Common Dissemination Channels**
  - **Integrated with Other User Information as required**

In conjunction with the User Applications Tiger Team and discussions with a broad spectrum of users, a space weather SAX was conducted. The SAX objectives were to capture and assess user insight on the utility of space weather information to their planning and operations in support of a broad range of National Security and civil missions and functions.

In the area of products, the SAX results indicated that an expert system translating space weather information into user impact terms and autonomous space weather updates for correction of the user systems is needed. Users also expressed a need for standardized, integrated products and a space weather expert point of contact to be available to fill special product requests and analyses. In addition to a requirement for significant improvement in space weather specification, reliable 4-6 hour and 24-hour forecasts and advisories are needed to support the mission planning cycle.

Operational systems require high confidence space weather models. For National Security architectures, this means going through rigorous verification, validation and accreditation processes.

Increased use of expert systems and tactical decision aids (TDAs) for mission planners and operators creates a need for space weather information (not data) to be smoothly integrated with CONOPS, contingency planning and standard situational awareness displays.

The civil community has the same need for "impact" specification and forecasts as the National Security community. However, the civil community relies on a network of value added resellers to provide user specific products that use space weather assets and data.

In summary, the ADT found:

- Products currently available to operators and planners are inadequate

- Most users need space weather information provided in terms of impacts and in formats that readily integrate into existing or planned systems

- Users need high confidence in space weather products for operational decisions and medium confidence for longer term plans

- Military users expect tailored space weather products while civil policy is to provide access to basic data and rely on third party product tailoring

C.2.7 Integrated Space Weather Center

To guide future investment, development and acquisition of space and space-related space weather capabilities, the NSSA recommends:

- **Evolve to an Integrated Space Weather Center capability to include:**
  - **Space Weather Expertise available for User Consultation and Support**
  - **A National Security Support Cell to produce Tailored Products**
  - **Back-up capability to provide support in the event of Natural Emergencies or Catastrophic Equipment Failures**

One important trade axis explored was distributed — centralized processing. Performance and cost were evaluated for architectures at the extremes. Centralized processing emerged as the better approach. In addition, the need for coupled computer-intensive models using consolidated global data drove the need for a highly capable central processing facility. This facility requires access to all data sources including space weather climatology and space weather impacts. Some users will require unprocessed data and their needs can be easily met with this centralized approach.

There is a high level of cooperation between the military and civil space weather centers including sharing data, models, and personnel. However, it is clear there is a potential for cost saving by evolving to an integrated space weather support capability. Because of the unique needs of some users, a National Security cell for processing classified data or providing classified products will be needed. Space weather center integration could compromise robustness, so a back-up center must be considered to reduce vulnerability to natural disasters and catastrophic equipment failure.

During the SAX, planners and operators also expressed a need for a space weather expert to be accessible to answer space weather questions and resolve issues.

In summary, the ADT found:

- The complexity of space weather models and forecasting will likely require a full time expert resource available to produce and evaluate products and interface with users.

- Centralized processing provides a single point of contact that is best for meeting most user needs

- An integrated space weather center with civil and joint military staffing along with back-up capabilities could improve efficiency and reduce costs in developing user products

- A National Security support cell is needed to focus on tailored products and classified support for DOD and Intelligence Community users

C.2.8 Space Weather Research and Development

To guide future investment, development and acquisition of space and space-related space weather capabilities, the NSSA recommends:

- **Provide a Robust space weather Research and Development Program to:**
  - **Develop and Implement the Improved Models**
  - **Provide options for further growth**
- **Continue to Leverage Research and Development Missions**
  - **Enhance Operational Products until Operational Systems are ready**
- **Develop and Implement Standardized Processes to rapidly and efficiently Transition R&D into needed Operational Products**

A technology assessment identified, characterized, and documented a technology foundation for post-2010 space weather capabilities. Three common threads were revealed.

First, multi-point measurements are vital for a complete picture of the environment. Space weather is currently starved for data essential to global specification, improved understanding, and better initialization and validation of forecast models (e.g., measurements from space Sun-Earth-Line sensors). Today, much of the data on space weather is limited to a certain geographical area or by resolution. Thus the current space weather architecture's ability to detect and mitigate space weather impacts is severely limited.

Second, basic research is an underpinning for better models. This research should focus on coupling process physics and space weather domain specific algorithms. High confidence forecasting can only be achieved with models integrated across the space weather domains (coupling from the Sun to the magnetosphere and through the ionosphere) that are verified and validated. The processes of how and when the Sun produces CMEs and the interaction between the magnetosphere and the ionosphere must be explored.

Third, new sensors and other supporting technologies are essential, but are largely driven by other than space weather needs. These supporting technologies include automated low cost spacecraft, low cost lightweight sensors, and advanced computing to run the complex space weather models. Promising sensors like solar flare and CME imagers will move us toward the ability to predict the impact to the Earth after detection of the event on the Sun. This lead-time will improve our forecasting ability. GPS occultation sensors and combined radiation and threat warning sensors that can be inexpensively deployed on a large number of satellites and significantly contribute to specification and forecasting of the magnetosphere and ionosphere. Lightweight payloads and low integration costs are the drivers for space weather sensors riding on other types of satellites.

More data and basic research are critical to model development and improvement. The ADT also validated the current practice of leveraging R&D missions (e.g., Advanced Composition Experiment (ACE)) to provide data to forecasters that would otherwise be unavailable. The use of these data increases forecast and specification confidence.

In summary, the ADT found:

- Space weather is a technically immature discipline and basic research is vital
- R&D sensors are a valuable data source and greatly benefit data-starved operations

- Flexible space weather architecture could allow easier transition of R&D to operations
- More focus on operational needs could improve R&D pay-offs
- Some R&D is ready for transition to operations now (e.g., Coronagraph, Compact Environmental Anomaly Sensor (CEASE), and GPS Occultation)
- R&D investment is key to reducing model development risk

C.2.9   Space Weather and Man-Made Effects Information Coordination

To guide future investment, development and acquisition of space and space-related space weather capabilities, the NSSA recommends:

- **Support the Space Control Protection Mission by providing timely Space Weather Information**
- **Incorporate the Operational Specification and Forecasting of Space Environmental Effects of Man-made (Primarily Nuclear) Events as a Mission into the Space Weather Architecture**

The ADT studied the relationship between man-made effects (MME) and space weather effects on the near-Earth environment. The spatial and temporal scales of most man-made effects are smaller than those of naturally occurring space weather phenomena, while high altitude nuclear explosion energy levels can be much higher than natural phenomena as well as other MME.

MME and space weather impacts are similar for high-energy photons (e.g., x-rays), pumped radiation belts, ionospheric disturbances, and aurora emissions/clutter. Space weather sensors can be used to trace MME, but they may not have the necessary dynamic range.

The Space Control mission requires the characterization of the natural environment to differentiate between outages caused by space weather or a hostile force. In many cases it is economical to field combined packages to provide threat warning, attack assessment, and space weather (e.g., CEASE). The space weather and MME physics models are similar, and the MME models require space weather information for initialization. In addition, nuclear detonation sensors can supply useful data to space weather modelers and forecasters. Thus sharing data and models between agencies is to be encouraged.

In summary, the ADT found:

- MME are physically similar to space weather effects, differing in that MME are more localized and have different energy levels
- Nuclear effects are the primary man-made threat to the space weather environment
- Users and models would benefit from spacecraft space environmental sensors
- Nuclear detection missions collect data that could benefit the space weather mission area
- Combining space weather and threat sensors would benefit the space control mission area

APPENDIX D

INFORMATION SOURCES AND POINTS OF CONTACT FOR NATIONAL SPACE WEATHER PROGRAM

Copies of *NSWP Strategic Plan* and *Implementation Plan*:

> Office of the Federal Coordinator for Meteorological Services
>   and Supporting Research (OFCM)
> Suite 1500, 8455 Colesville Road
> Silver Spring, MD 20910
> Phone: 301-427-2002   DSN: 851-1460
> Fax: 301-427-2007
> E-mail: OFCM.mail@noaa.gov

On-Line Copies of Plans:

> Use your web browser to navigate to http://www.ofcm.gov/ then follow
> links to on-line publications.

Key Members--Committee for Space Weather:

> **Co-Chair, National Science Foundation:**
>   Dr. Richard Behnke
>   Director, Upper Atmospheric Research Program
>   National Science Foundation
>   4201 Wilson Boulevard Room 790
>   Arlington, VA 22230
>   Phone: 703-306-1518   Fax: 703-306-0849
>   E-mail: rbehnke@nsf.gov

> **Co-Chair, Department of Commerce:**
>   Dr. Ernest Hildner
>   Director, Space Environment Center
>   National Oceanic and Atmospheric Administration
>   325 Broadway
>   Boulder, CO 80303
>   Phone: 303-497-3311   Fax: 303-497-3645
>   E-mail: ehildner@sec.noaa.gov

Co-Chair, Department of Defense:
Colonel Michael A. Neyland
HQ USAF/XOW
1490 Air Force Pentagon
Washington, DC 20330-1490
Phone: 703-614-7291   DSN: 224-7291
Fax: 703-614-0055
E-mail: michael.neyland@pentagon.af.mil

Executive Secretary (Office of the Federal Coordinator for Meteorology):
Lieutenant Colonel Michael R. Babcock until July 31, 2000
Lieutenant Colonel Mark Welshinger after July 31, 2000
(see OFCM contact information above)

Department of Energy:
Mr. Gerald Maxwell
Department Of Energy
Office of Nonproliferation and National Security
NN-20, Room GH-068
1000 Independence Ave
Washington, DC 20585
Phone: 202-586-5539
E-mail: jerry.maxwell@hq.doe.gov

Department of the Interior:
Dr. Arthur Green, Jr.
U.S. Geological Survey
MS-968
Denver Federal Center
Denver, CO 80225
Phone: 303-273-8482   Fax: 303-273-8450
E-mail: awgreen@gldfs.cr.usgs.gov

National Aeronautics and Space Administration:
Dr. George Withbroe
NASA Headquarters
Research Program Management Division
Code SR
Washington, DC 20546
Phone: 202-358-2150   Fax: 202-358-3096
E-mail: george.withbroe@hq.nasa.gov

Department of Transportation:
Mr. Erwin Williams
Federal Aviation Administration Headquarters
Weather Standards Division
Room 8320D
400 7th Street, S.W.
Washington, DC 20590
Phone: 202-366-4629   Fax: 202-366-5549
E-mail: erwin.ctr.williams@faa.gov

Stating Requirements for Space Weather Support:

Civilian Requirements:
Space Environment Center
325 Broadway
Boulder, CO 80303
Phone: 303-497-5687   Fax: 303-497-7392
e-mail: gheckman@sec.noaa.gov

Military Requirements:
Commander, 55th Space Weather Squadron
400 O'Malley Avenue Suite 60
Schriever AFB, CO 80912-4060
Phone: 719-567-6206   DSN: 560-6206   Fax: 719-567-6219

Joint Modeling and Simulation Programs:
Office Chief, Air and Space Natural Environment
Modeling and Simulation Executive Agent
151 Patton Avenue, Rm 120
Asheville, NC 28801-5002
Phone: 828-271-4209/4322 DSN: 697-9016   Fax: 828-271-4324
E-mail: asne@afccc.af.mil

Contacting the Community Coordinated Modeling Center (CCMC):

Dr. Michael Hesse, Director
Code 696
NASA/Goddard Space Flight Center
Greenbelt, MD 20771
Phone: (301)-286-8224   Fax: (301)-286-1648
E-mail: hesse@gsfc.nasa.gov

APPENDIX E

ABBREVIATIONS AND ACRONYMS

ACE	Advanced Composition Explorer
AE	auroral electrojet
AFB	Air Force Base
AFOSR	Air Force Office of Scientific Research
AFRL	Air Force Research Laboratory
AFSCN	Air Force Satellite Control Network
AFWA	Air Force Weather Agency
AF/XOW	Headquarters U.S. Air Force, Directorate of Weather
AGO	Automated Geophysical Observatory
AGU	American Geophysical Union
Ap	Planetary A index of geomagnetic activity
ARGOS	Advanced Research and Global Observing Satellite
AU	astronomical unit
CCMC	Community Coordinated Modeling Center
CEDAR	Coupling Energetics and Dynamics of Atmospheric Regions
CME	coronal mass ejection
CRADA	Cooperative Research and Development Agreement
CRL	Communications Research Laboratory, Tokyo, Japan
CRRES	Combined Release and Radiation Effects Satellite
CSEF	Committee for Space Environment Forecasting (OFCM)
CSTR	Committee on Solar Terrestrial Research (National Academy of Sciences)
CSW	Committee for Space Weather (OFCM)
CTIM	coupled thermospheric-ionospheric model
CY	calendar year
DISS	Digital Ionospheric Sounding System
DMSO	Defense Modeling and Simulation Office
DMSP	Defense Meteorological Satellite Program
DOC	Department of Commerce
DOD	Department of Defense
DOE	Department of Energy
DOI	Department of Interior
DOT	Department of Transportation
DPU	Data Processing Unit
DSN	Deep Space Network
Dst	magnetic disturbance storm time index
E-field	electric field
EISCAT	European Incoherent Scatter Radar
ESA	European Space Agency
EUV	extreme ultraviolet
eV	electron-volt
EVA	Extra-Vehicular Activity

FAA	Federal Aviation Administration
FAST	Fast Auroral Snapshot
FCMSSR	Federal Committee for Meteorological Services and Supporting Research (OFCM)
foF2	Maximum ordinary mode radiowave frequency capable of reflection from the F2 region of the ionosphere
FY	fiscal year
GEC	global electrodynamics connection
GEM	Geospace Environment Modeling Program
GEO	Geosynchronous orbit
GEOSpace	Workstation software suite of space weather and related applications
GGS	Global Geospace Science Program
GHz	gigahertz
GIC	geomagnetically induced currents
GLO	Arizona Airglow Experiment
GOES	Geostationary Operational Environmental Satellite
GPS	Global Positioning System
GPS/MET	Global Positioning System Meteorological Sounding Experiment
HEO	highly elliptical Earth orbit
HESSI	High Energy Solar Spectroscopic Imager
HF	high frequency
ICMSSR	Interdepartmental Committee for Meteorological Services and Supporting Research (OFCM)
IMAGE	Imager for Magnetopause to Aurora Global Exploration
IMF	interplanetary magnetic field
IPS	interplanetary scintillation
IR	infrared
IRI	International Reference Ionosphere
ISES	International Space Environment Service
ISOON	Improved Solar Observing Optical Network
ISTP	International Solar-Terrestrial Physics Program
IUWDS	International URSIgram and World Days Service
JHU/APL	The Johns Hopkins University/Applied Physics Laboratory
K-12	kindergarten through 12th grade
keV	hundred electron-volts
kHz	kilohertz
km	kilometers
Kp	Planetary K index of geomagnetic activity
L1	Lagrangian point
LAN	local area network
LEO	low Earth orbit
LORAN	Long-range Radio Navigation
LTE	local thermodynamic equilibrium
LWS	Living with a Star Program (NASA)
MeV	million electron volts

MHD	magnetohydrodynamics
MIT	Massachusetts Institute of Technology
MLTI	mesosphere and lower thermosphere/ionosphere
MSFM	Magnetospheric Specification and Forecast Model
MSIS	Mass Spectrometer and Incoherent Scatter model
MSM	Magnetospheric Specification Model
MSX	Midcourse Space Experiment
NAS	National Academy of Sciences
NASA	National Aeronautics and Space Administration
NGDC	National Geophysical Data Center
NmF2	peak F-region electron density
NOAA	National Oceanic and Atmospheric Administration
NPOESS	National Polar-orbiting Operational Environmental Satellite System
NRL	Naval Research Laboratory
NSF	National Science Foundation
NSSA	National Security Space Architect
NSSDC	National Space Science Data Center
NSWP	National Space Weather Program
NSWPC	National Space Weather Program Council (OFCM)
OFCM	Office of the Federal Coordinator for Meteorology
OMB	Office of Management and Budget
PI	power index
PL	Phillips Laboratory
POES	Polar Orbiting Environmental Satellite
PRISM	Parameterized Realtime Ionospheric Specification Model
RAL	Rutherford Appleton Laboratory, United Kingdom
RAO	Relocatable Atmospheric Observatory
R_E	Earth radii
RPC	Rapid Prototyping Center
R_S	solar radii
RSTN	Radio Solar Telescope Network
RTSW	Real Time Solar Wind
SAMPEX	Solar Anomalous Magnetospheric Particle Experiment
SAX	Space Architect Exercise
SBIR	Small Business Innovative Research
s/c	spacecraft
SCINDA	Scintillation Network Decision Aid
SEC	Space Environment Center (NOAA) or NASA's Sun-Earth Connections Program
SEM	Space Environment Monitor
SEON	Solar Electro-Optical Observing Network
SEP	solar energetic particles
SHINE	Solar Heliospheric and Interplanetary Environment Program
SMC	Space and Missile Systems Center, Air Force Materiel Command
SMEI	Solar Mass Ejection Imager

SOHO	Solar and Heliospheric Observatory
SRBL	Solar Radio Burst Locator
SSI	Space Sciences Institute
SST	supersonic transport
STOA	shock time of arrival
SuperDARN	Super Dual Auroral Radar Network
SWRN	Space Weather Research Network
SXI	Solar X-ray Imager
TEC	total electron content
TIEGCM	Thermosphere-Ionosphere Electrodynamics General Circulation Model
TIMED	Thermosphere-Ionosphere-Magnetosphere Energetics and Dynamics Mission
TIROS	Television and Infrared Observation Satellite
TRACE	Transitions Region and Coronal Explorer
T^2	technology transfer
UARS	Upper Atmosphere Research Satellite
UCLA	University of California at Los Angeles
UHF	ultrahigh frequency
UML	University of Massachusetts at Lowell
URL	World Wide Web Universal Resource Locator
USAF	United States Air Force
USGS	United States Geological Survey
UV	ultraviolet
VHF	very high frequency
W	watt
WAAS	Wide Area Augmentation System
WAN	wide area network
WBMOD	Wideband Model
WG/NSWP	Working Group for the National Space Weather Program (OFCM)
3D	three-dimensional
55 SXWS	55th Space Weather Squadron (USAF)

Contributors

National Science Foundation
Dr. Richard Behnke
Dr. Robert Robinson
Dr. Kile Baker
Dr. Sunanda Basu

National Oceanic and Atmospheric Administration
Mr. Gary Heckman
Ms. Amy Holman

Department of Defense
Lt Col Michael Bonadonna, USAF
Maj Paul Bellaire, USAF
Capt Riley Jay, USAF
Dr. Robert McCoy, USN

National Aeronautics and Space Administration
Dr. Lawrence Zanetti
Dr. Michael Hesse

Department of Transportation
Lt Col Erwin Williams, USAF

**Office of the Federal Coordinator for Meteorological Services
and Supporting Research**
Lt Col Michael Babcock, USAF
Editor

Graphic courtesy of National Geophysical Data Center (NGDC), NOAA, Boulder, CO

**Office of the Federal Coordinator
for
Meteorological Services and Supporting Research**

**8455 Colesville Road, Suite 1500
Silver Spring, Maryland 20910**

**Telephone: (301) 427-2002
Fax: (301) 427-2007
http://www.ofcm.gov/**